PRINTING PROPS

A Beginner's Guide to 3D Printing for Cosplay and Replicas

By

Christopher Maddalena

For Carla and Quentin.

Printing Props by Christopher Maddalena

Layout by Christopher Maddalena
Illustrations by Christopher Maddalena
Photography by Christopher Maddalena
Edited by Christopher Maddalena, Jeff Dimmock, and Mike Brown

Contents

Opening Thoughts

I started dabbling in woodworking in 2006 but never considered myself a maker. I was just someone who made a decent cutting board or shoe rack when I wasn't working as a consultant or playing video games. Through a quirk of fate, video games led me to Whiskey Media, which would create the websites Tested and Giant Bomb. When the Whiskey Media crew put on their second annual Big Live Live Show, a big variety show of sorts, they invited Harrison Krix from Volpin Props to show his outstanding Mass Effect and Bioshock props. Those props were things I never thought I'd be capable of making at my home. Harrison later returned to share his build of the Needler gun from Halo. Hearing how he used molds to create it translated in my head as, "You need special equipment. This isn't something for an amateur."

In 2012, Whalerock Industries purchased Tested, and it became what is now Adam Savage's Tested. Adam Savage appearing on Tested reintroduced him to me as a maker and a creative. Before that, I was a fan of Mythbusters, his TV show on Discovery Channel with Jamie Hyneman, but I never saw him talking about his creative process until Tested. Adam Savage became a powerful influence on me. In particular, his positive attitude towards trying new things and being proud of what you have made.

In 2016, Seattle-based maker Bill Doran of Punished Props released a video on YouTube titled "Cheap & Fast Han Solo DL-44 Blaster Tutorial." In this video, Bill walked through the process of repainting a $12 bright orange toy gun made by the costuming company Ruby's to make it look like a beat-up blaster out of Star Wars. I was blown away by what he proved was possible with some rattle cans and mustard (yes, the condiment). I was even more excited when I followed along and achieved results that at least impressed friends and family at a Halloween party.

I kept following build updates from Tested, Punished Props, and Volpin Props. They introduced me to more makers that showed me the fun and creative ways they achieved mind-blowing results. Two years later, I revisited that blaster and made a more screen-accurate DL-44 using 3D printed parts, silicone molds, and cold cast resin. I added a 3D printer to my basement shop in 2017 after watching many more Punished Props videos and being inspired by Bill's use of 3D design and printed materials in his projects.

I have seen people say, "3D printing doesn't count as making." I once felt that way, that 3D printing something was a shortcut or a cheat. I look back now and can see how that sapped the joy from my finished pieces. I felt like my props were somehow lesser than others. That thought was wrong, and that was one of the inspirations for this book.

I started writing this book in January 2020. I had a draft of something like a technical manual when many things changed in the world over the next few months. There's no reason to record specifics here, but I will say those changes made me feel as if the book needed to change, too. 3D printing and making had brought me so much joy that I wanted that reflected in the book. I wanted to write something that recreated passionate discussions with friends. I also wanted to showcase members of the maker community and their designs that had inspired me. I threw out dozens of pages and set out to make something conversational, fun, and enjoyable to read.

I hope I achieved that goal and that this book will help at least a few people get started 3D printing and find a similar joy. It can look messy, expensive, and complicated from the outside. Sometimes it's all of those things, but it gets more comfortable with practice and understanding. May this book help you through some of the rough spots and moving towards a lifetime of rewarding and joyful experiences.

I cannot end this introduction without mentioning some of the makers who have helped and inspired me. I am writing this book because of these people:

- Bill and Britt Doran (Punished Props, *https://www.punishedprops.com/*) – Still a continuous source of inspiration to me, and they have created a unique and supportive community of makers in their "Proptarts" social media group.
- Simone Giertz (*http://www.simonegiertz.com/*) – Simone animates me with her creative projects, like a coffee table made of matchsticks. She is an amazing engineer and creative.
- Svetlana and Benni Quindt (Kamui Cosplay, https://www.kamuicosplay.com/) – Creators of some genuinely incredible cosplay outfits and high energy videos. Just try not to smile as Svetlana excitedly presents their latest project updates on YouTube. Buy her books!
- Evan and Katelyn (*https://evanandkatelyn.com/*) – These two have some of the most fun videos that continuously provoke me to look at my everyday surroundings and make something extraordinary to replace the mundane.
- Harrison Krix (Volpin Props, *http://www.volpinprops.com/*) – The source of so many jaw-dropping props and trophies seen at gaming competitions and conventions. You owe it to yourself to check out the Volpin Props portfolio.
- Adam Savage (Tested.com, *https://www.tested.com/*) – A maker role model and, in his own words, a high functioning hoarder of some of the coolest objects he has created and collected over the years.

These makers, and many others, have laid a foundation with their podcasts, books, videos, and articles–by being who they are and reveling in things they love–that has made it easier for people like me to learn and create something that brings myself and others joy. This maker movement can only lead to the next generation, like my young son, getting involved and learning at an earlier age and growing up feeling empowered to make awesome things.

To these makers and others: Thank You. What you do means a lot, and I certainly haven't teared up while trying to write some of these words.

CHRISTOPHER MADDALENA

NOV 2020

How to Use This Book

This book is intended to be something you can read from cover to cover and use as a reference while planning how you might incorporate 3D printing into your next prop or cosplay project. Sections begin by introducing basic 3D printing fundamentals and advance towards topics like designing custom models.

This book is not a technical manual. There is no way for this book to provide information like specific printer or material settings. Printer configurations change based on many variables (including your own personal preferences and environment), so technical discussions will be high-level. Where possible, sections provide references to outside resources where you can learn more.

Each section features a photo of a prop or model I really enjoyed making. Many of the 3D models came from online shops–like the one over there! Every image has a description that tells you what it is, the source material, a brief description of how I made it, and who designed the model.

The last section has lists of references and links to learn more about some of the advanced topics and find the incredible models that appear throughout the book.

One of this book's goals is celebrating makers who have shared their remarkable talents and creations with the community. Please, take some time to check out their models and shops.

Finally, visit *printingprops.com* for a growing collection of additional links, resources, and design files for some of my custom models featured in these pages.

Headpiece of the Staff of Ra

Raiders of the Lost Ark

Model by Digital Reflection[1]

Printed in resin and finished with Alclad II lacquers and oil paints

Gem printed in red transparent resin, in two halves, with epoxy in-between

Safety First

A 3D printer is a power tool, and like any power tool, it requires care and respect. If you are complacent or careless, a printer can cause you serious injury or fall into disrepair. You may also be exposed to nasty dust and dangerous fumes while working with your 3D prints.

Throughout the book, you will see some **blue notes** that call-out additional safety tips and warnings.

Follow these recommendations to stay safe!

Read Your Manuals

Before you do anything, read your printer's manual and learn how it works and how to maintain it. Do the same for all of your other tools, glues, and solvents to understand how to use and store them.

3D printers require calibration and maintenance to keep them operating at peak performance. Belts and screws will loosen, gears will collect debris, wires may fray, and nozzles will clog. If left unattended, these issues will degrade your print quality and create safety hazards. Read the manuals and work towards understanding how your printer works so you can maintain it, operate it safely, and get top-notch prints.

Place Printer in a Safe Place

Fused Deposition Modeling (FDM) printers use "open-loop" systems that create opportunities for pinch injuries. Printers are dumb and only do what the code tells them to do. If a finger gets between a stepper motor and the printer's frame, you can be seriously injured. Most printers cannot detect the pinch and will not stop trying to move.

An FDM printer's "hot end" gets hot enough to cause severe burns on contact. Heated print surfaces are large areas that reach upwards of 110°C. It is easy for an adult to burn themselves accidentally, so this is a big concern for curious children.

Stereolithography (SLA) printers hold vats of liquid resin. The resin is toxic, a skin irritant, and capable of creating incredible messes. You do not want young children and pets knocking over the printer or getting into the resin.

Further, all 3D printers are rather large and heavy objects. A child would be seriously hurt if they were to pull the printer down on top of themselves.

Place printers out of reach of children. Consider building an enclosure for any FDM printers. An enclosure keeps dust and pet hair off the printer, prepares you for successfully printing ABS, and acts as a barrier against little fingers getting near pinch points and hot metal.

Fire Prevention

The "hot end" of an FDM printer parts gets hot enough to start fires, so either keep the printer near a smoke detector or place a smoke detector next to the printer. If you build an enclosure, it's a good idea to install a smoke detector inside the enclosure to monitor the printer.

Never leave your printer entirely unattended. You will inevitably print something that requires many hours and can't be supervised. Just don't start an 18-hour print and leave home for the weekend. Remain nearby so you can respond to a print failure, or a worse situation.

During maintenance, check for frayed wires. If the printer's movements pinch or rub a wire, it may become frayed over time. A wiring failure could harm your printer or cause a more severe problem like a fire.

As a precaution, keep flammable materials a safe distance away from the printer. Bottles of acetone and Isopropyl Alcohol (IPA) could help a small fire spread rapidly.

Shock Prevention

FDM printers tend to have all metal frames that conduct electricity very well. If the printer is not grounded, wiring issues may cause electricity to flow through the frame. Touching the frame could give you a nasty shock.

Always ground your 3D printer to avoid any surprise shocks. If you notice anything wrong, or even a slight tingling when near the frame, unplug the printer before investigating.

Wear a Respirator

Before you reach for resin, epoxy, acetone, or sandpaper, put on a face mask. Sanding kicks up a lot of dust that you do not want to breathe, so wear a dust mask when you start a sanding session. Likewise, put on a respirator before using acetone or toxic glues.

3M's 6001 organic vapor filter is an excellent all-around choice for your shop. It will protect you against fumes from acetone and the glues (e.g., contact cement) you are likely to use with 3D prints, foam, and other common cosplay materials.

Write a date on your filters, so you know when you put them into use. Change them regularly! 3M recommends changing filters as soon as you can smell a hint of the fumes through your mask. Filters also have a shelf life of about six months once you open the package.

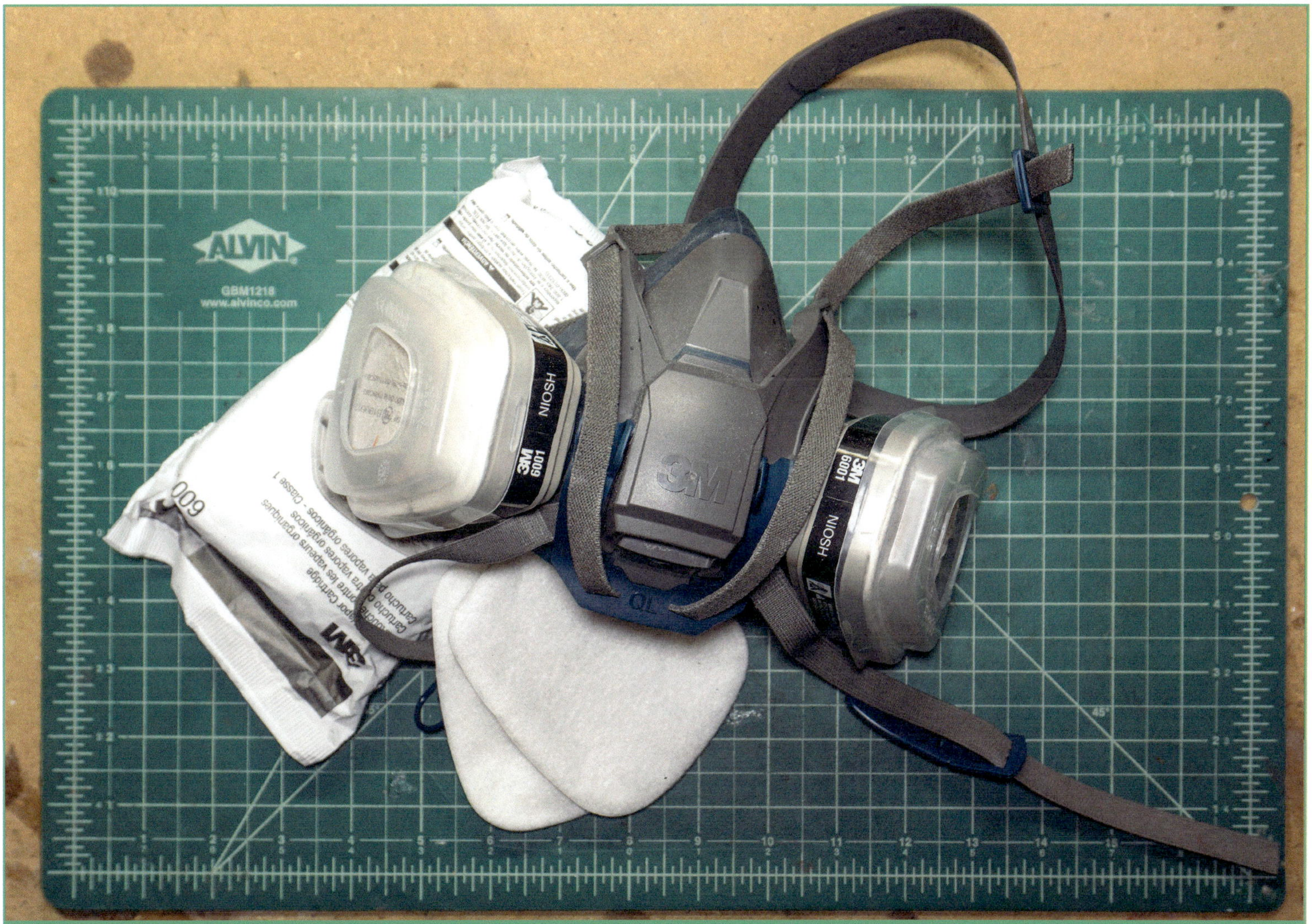

3M 6500 Respirator with Spare 6001 Filters

Tip

Check out 3M's 6500 series of half facepiece respirators. These feature 3M's "quick latch drop-down mechanism" that makes it possible to pull a latch on the mask's front and drop it down to your chin. Very handy for taking a drink of coffee and talking to people while working.

Wear Gloves and Safety Glasses

Removing a finished print from the bed may involve using a sharpened scraper tool. Using a scraper is less common now, but it is occasionally necessary, and many printers still ship with a scraper in the box. You can badly injure yourself if you slip, especially when you are pushing towards your supporting hand. Wear gloves to protect your hands from these mishaps.

Likewise, little plastic pieces tend to fly through the air when removing small models or support material. Wearing safety glasses may seem like overkill, but you do not want to experience having a shard of support material stuck in your eye. Shrapnel goes flying more than you might expect.

Speaking from experience, the moment you decide not to put on the glasses is when the support material will spring into your eye. It hurts! Be safe.

Finally, always wear gloves when handling resin and other chemicals you may use for post-processing your prints. Keep some disposable nitrile gloves in stock in your shop to keep that stuff off your hands. Cheap gardening gloves with rubber palms and fingers–like the orange pair below–can be a great alternative to nitrile gloves. They are reusable, and the elastic backs help your hands stay cooler and less sweaty; however, the rubber is usually textured for an improved grip, which can leaves marks on your prop if you're wearing them for painting.

A Small Sample of My Collection of Safety Gear

Tip

Look for a good pair of wrap-around safety goggles like these Dewalt DPG82-11C goggles. They aren't necessary for 3D printing, but they're fantastic for protecting your eyes while using rotary tools, band saws, and other tools you're likely to use for your projects.

Banzai Bill

Super Mario World

Model by Martin Moore (Forge 3D)[2]

Printed in matte PLA with no finishing work

Getting Started

This section covers the basics of 3D printing as it relates to cosplay and props. We'll begin by covering the common printing technologies, how they affect your decisions when printing a part for a prop, and where you can get some models to help you get started.

Fused Deposition Modeling

Fused Deposition Modeling (FDM) is the most ubiquitous 3D printing technology. FDM works by using movements on the X, Y, and Z axes (horizontal left/right, horizontal forward/back, and vertical) to lay down material onto a print platform (also called the print bed). There are various materials used with FDM (discussed later). Still, most 3D printing hobbyists use materials that fall under the general category of *thermoplastics*. When heated to a high enough temperature, these plastics become pliable.

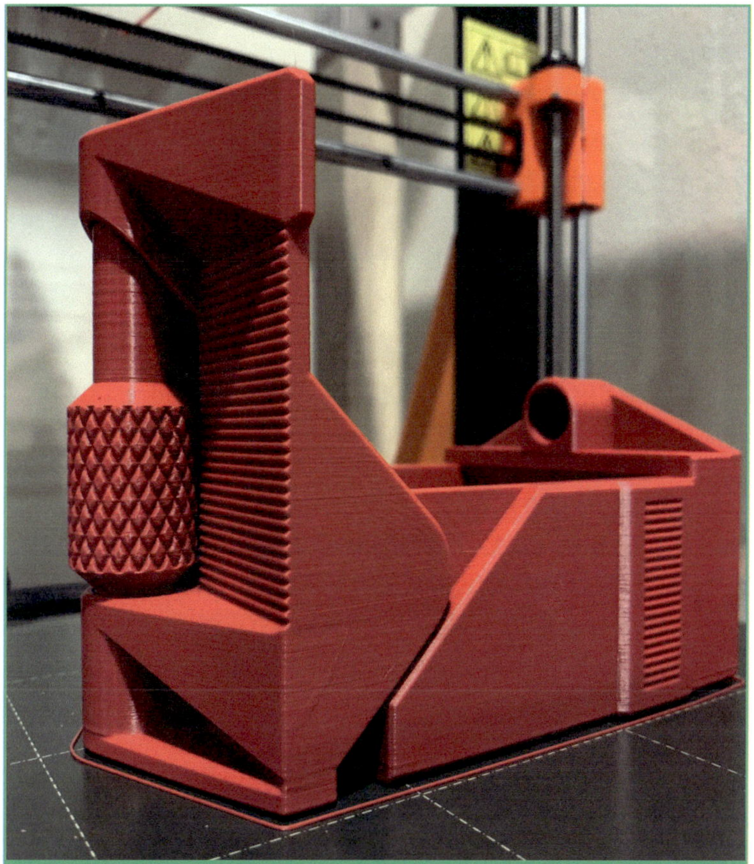

Box Printed as One Piece on an FDM Printer

FDM printers can print all kinds of materials, even different types of metal! Although, printing metal is still the realm of commercial printers (for now).

The printer heats the material just enough so it flows smoothly through a nozzle and onto the print platform. This material is called *filament* because of its thread-like nature. It is commonly 1.75mm wide but also comes in 2.85mm for larger printers. The printer's extruder motor feeds the filament into a *heating block* and *nozzle*–also called the *hot end* of the extruder.

Models print in layers. When the printer completes a layer, it moves the hot end up a fraction of a millimeter and begins printing the next layer.

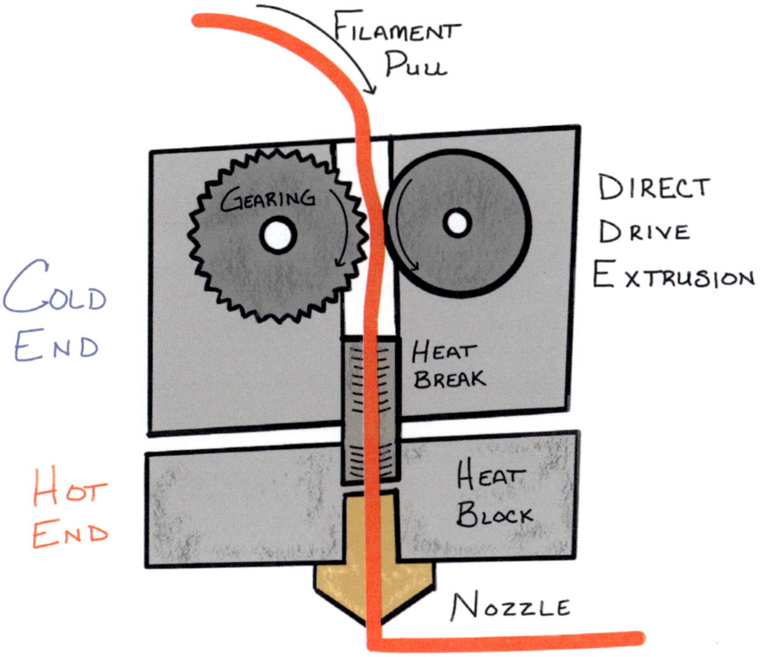

Note

FDM is the most common name for this technology, but FDM is technically a trademark name for Fused Filament Fabrication (FFF). If you see FFF used to describe a printer instead of FDM, just know they are the same.

Stereolithography

Stereolithography (SLA) is a printing technology that has only recently become affordable to hobbyists, but it is older than FDM. Online communities and articles often refer to SLA printing simply as "resin printing" because it uses liquid photopolymer resins instead of spools of solid filaments.

There are different variations of SLA. This book deals primarily with *inverted* SLA, the most common variation used by consumer machines. SLA printers use the same basic concepts of printing a model in layers but are different in nearly every other way.

These printers have a tank of resin below the printing platform with a thin transparent film–usually *fluorinated ethylene propylene* (FEP) film–on the bottom. The printer shines an ultraviolet (UV) light through the film. The resin cures (hardens) when exposed to UV light, so the printer lowers its platform to the bottom of the tank and selectively exposes the resin to UV light to complete a layer. Then, similar to FDM printing, the printer adjusts the platform to be slightly higher and cures the next layer.

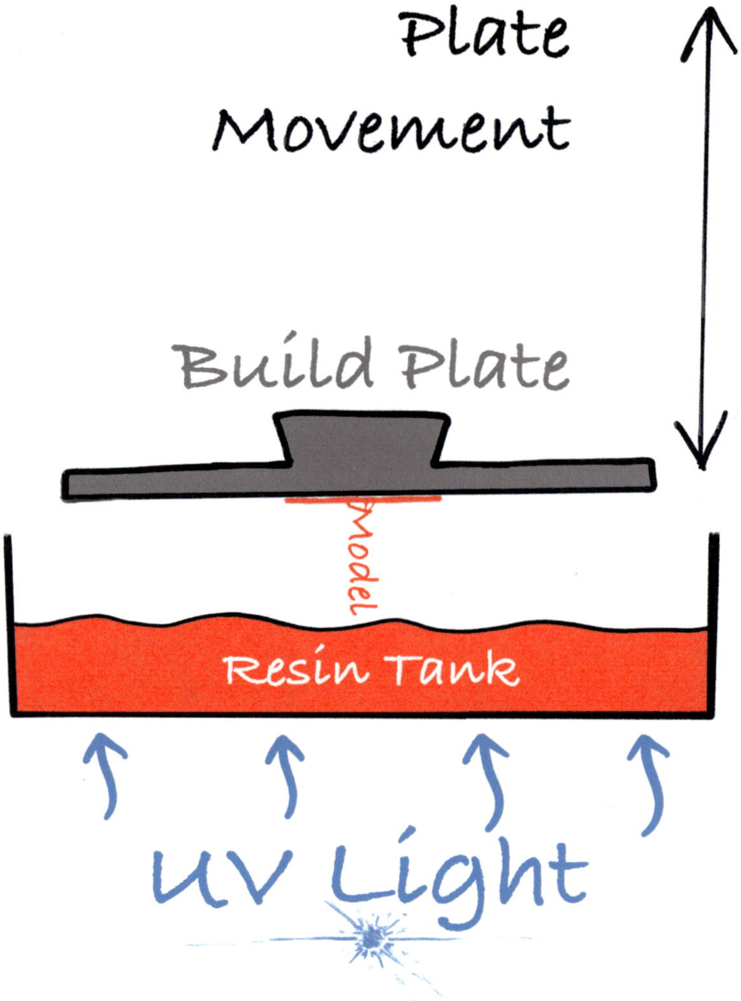

This process is exact, often involving lasers (e.g., the Formlabs Form 3 printer) to cure the resin so that prints can be incredibly detailed. A newer form of SLA printing, Mass SLA (MSLA), uses Liquid Crystal Display (LCD) screens to shine an entire layer's worth of light onto the resin all at once. MSLA printers (e.g., the Prusa SL1 printer) use high-resolution displays positioned below the tanks to do this.

One printing technology does not replace the other. There are advantages and disadvantages to FDM and SLA, especially when considering them from a prop maker or cosplayer's perspective. These differences will become apparent in the following sections on materials.

Note

SLA printers can produce prints with incredible levels of detail that look terrific with little to no post-processing work.

I printed this bust of *The Mandalorian* by David Östman[3] on the Prusa SL1 and took this photo before I completed any sanding or touch-up work.

FDM or SLA for Cosplay

As mentioned earlier, the FDM and SLA printing technologies each offer advantages and disadvantages. For maximum flexibility in your shop, you want both. However, it is wise to start with just one, see how you like it, and master it before moving forward with a second printer.

FDM printers are the most flexible and user-friendly and cover the broadest range of applications. These printers can create durable plastic pieces with excellent levels of detail. FDM printers can also print helpful tools that might aid you during a project. For example, a custom 3D printed box works well as a casing for making silicone molds. FDM is the technology that brings the most to the table for general 3D printing fun, props, and cosplay.

> **Tip**
>
> When you start out, stick with one type of printer and one material to reduce variables. Limiting your variables will make troubleshooting and learning much easier.

If you intend to print only smaller objects with high levels of detail, like jewelry or tabletop miniatures, an SLA printer might be a better choice for you. Just remember, the resin prints are brittle and will not work well for items that need to snap together. There are "strong" resins intended to be more durable than standard resin. Even these have their limits.

Further, larger SLA prints are typically hollow to reduce the weight that might otherwise pull the model off the build plate. If you want to add more weight to your larger prints, you will need to fill them with something like an epoxy or urethane resin.

When it comes to selecting a printer, there are hundreds of options. The selection process can be overwhelming and difficult, but don't worry too much about picking "the right one." Just about any well reviewed budget printer will work well for props and cosplay. The more inexpensive printers tend to print rougher models, but you will sand and paint most of the things you print (for props, anyway).

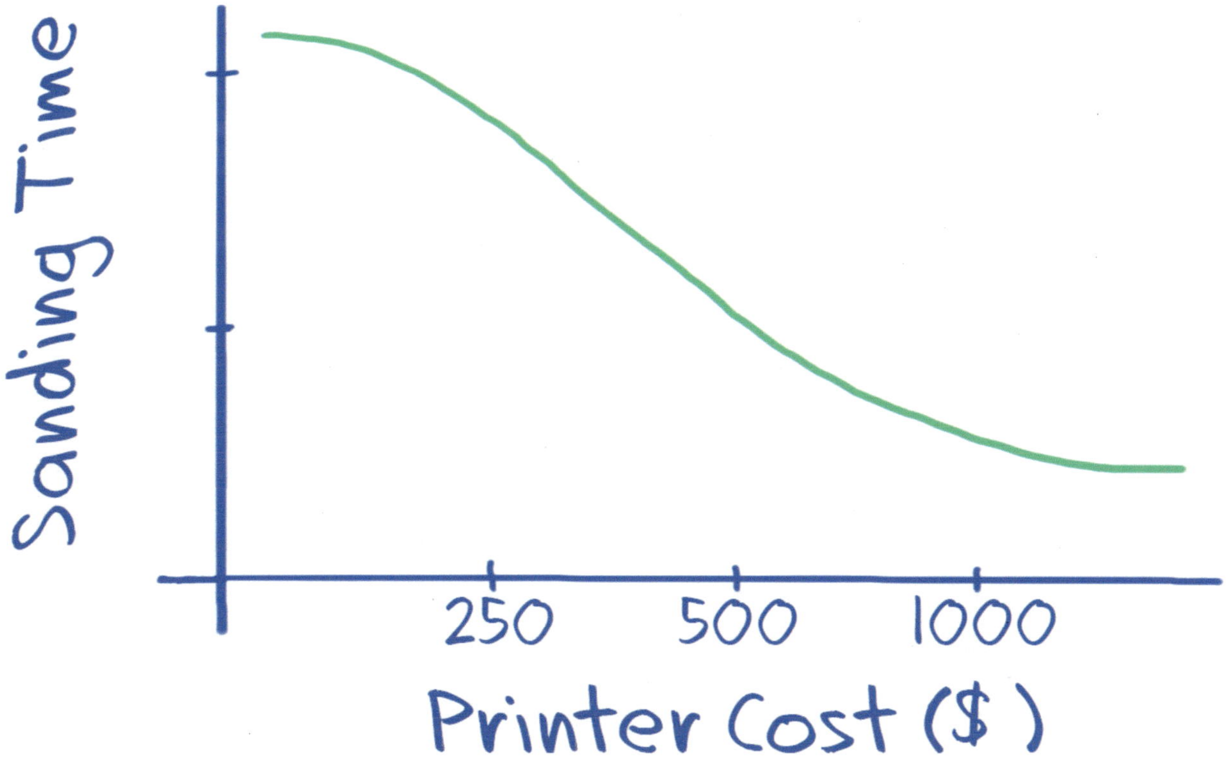

In general, the more you spend on a printer, the less time you will spend checking-in on prints and sanding your finished models. Pick one that matches your budget and complete your purchase with confidence.

After materials, a printer's build volume may be the second most impactful variable. Build volume refers to how large the print bed is and the Z-axis clearance (the maximum height for printing). This applies to both FDM and SLA printers.

FDM printers commonly have build volumes around 200 x 200 x 180mm (7.87 x 7.87 x 7.09in) to 220 x 220 x 250mm (8.66 x 8.66 x 9.84in). Sizes vary a great deal, but many popular printers (especially entry level or budget-frindly printers) have build volumes in this range.

The less expensive FDM printers and SLA printers have much smaller build volumes. The Elegoo Mars SLA printer has a build volume of just 120 x 68 x 155mm, and the budget Wanhao Duplicator i3 Mini FDM printer has a build volume of 120 x 120 x 120mm.

It can be tricky to figure out how big you might need your build plate to be. These builds volumes can sound bigger than they are. Adding just 10 or 20mm to a 200 x 200mm print bed can mean not having to cut or scale down a model.

For reference, this book is printed on 8.5 x 11" paper, which is larger than a 220 x 220mm print bed. This photo of the Prusa SL1 build plate is 1:1 scale and approximates the build volume of some of the most popular MSLA printers.

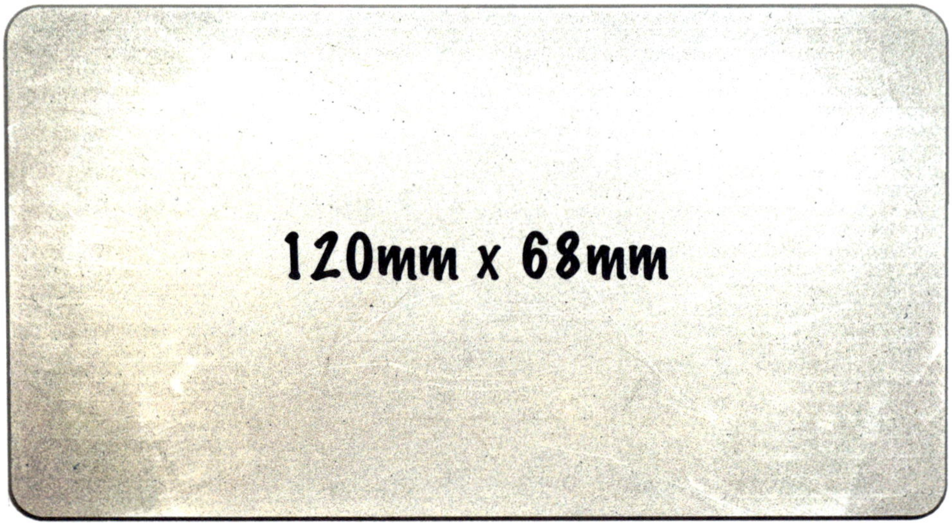

If printing larger objects in one piece is important to you, an FDM printer is more likely to offer that sort of volume. That said, there are new SLA printers with much larger print beds that are now within reach to some consumers, like the Peoply Phenom with its 276 x 155 x 400mm bed. The Phenom can print large objects in resin; however, it sacrifices some bells and whistles in favor of that larger build platform to get the total price down to $1,799 (as of 2020). That is relatively inexpensive for a large SLA printer, but still beyond most hobby budgets. A large build volume increases the printer's cost, especially on an SLA printer.

I used a Wanhao i3 Duplicator with a 200 x 200mm bed as my primary FDM printer for about two and a half years. This was ample space when I got started. Some models posed challenges, but models can be cut or cleverly oriented to make them work with a smaller print bed. This just adds the step of having to glue two pieces together.

Cutting a model isn't such a bad thing. If done smartly, reassembly can be easy, and seams will be hidden as part of your finishing process. We'll discuss editing models in a later section.

Slicing 3D Models

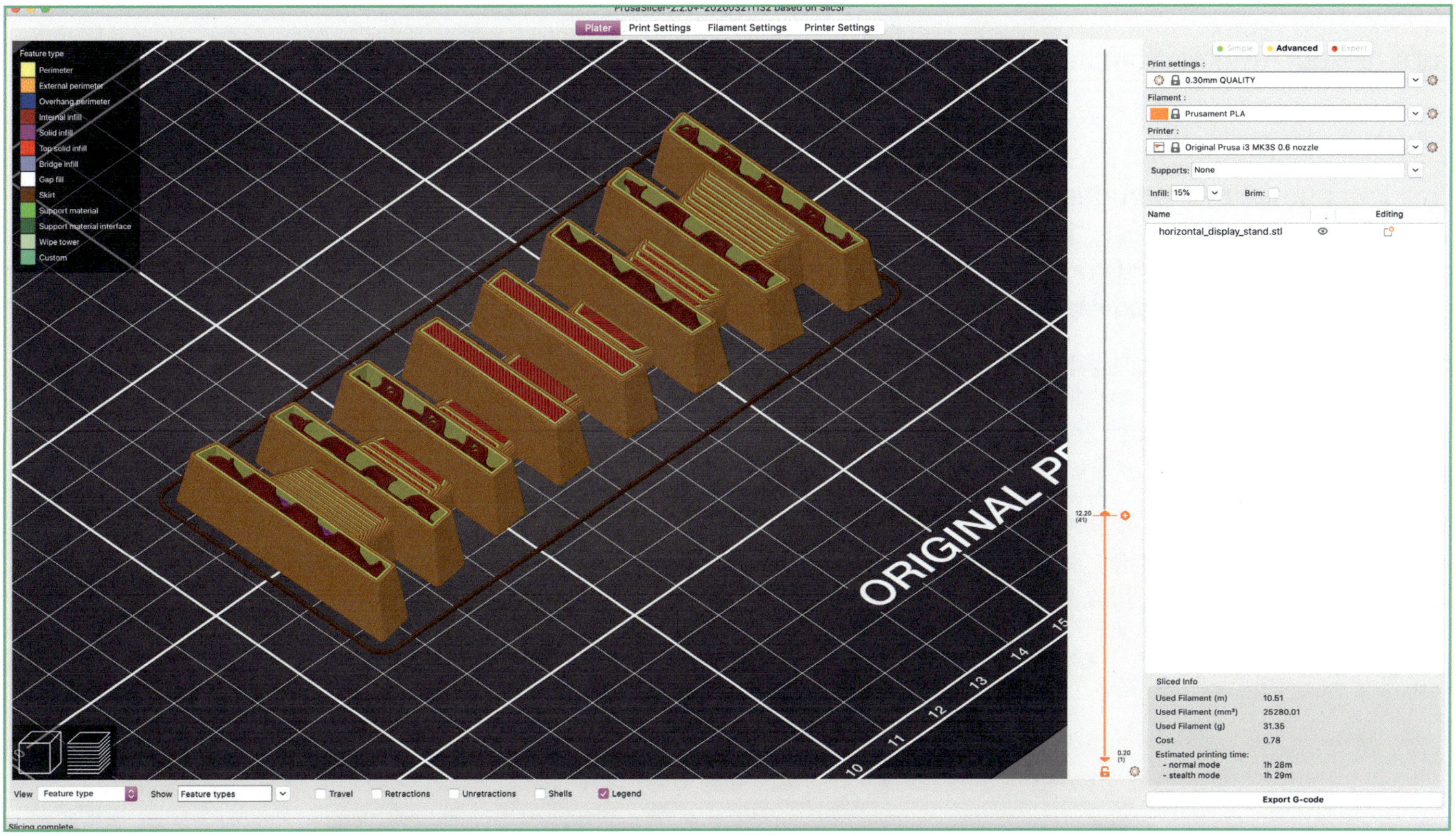

Preparing a 3D model for printing is called *slicing*. Slicing a model requires a class of software called *slicers*. These slicers take a model, cut it into layers based on your settings, and produce a file for your printer. For FDM printers, these files are Geometric Code, or G-code, files. G-code is a programming language that controls the movements of 3D printers and Computer Numerical Control (CNC) machines. The Massachusetts Institute of Technology developed G-code in the 1950's.

A slicer's exported *.gcode file provides instructions for every movement and action the printer performs.

For example, Prusa Research's printers add this snippet of G-code at the start of every print.

```
G28 W                        ; home all without mesh bed level
G80                          ; mesh bed leveling
G1 Y-3.0 F1000.0             ; go outside print area
G92 E0.0
G1 X60.0 E9.0 F1000.0        ; intro line
G1 X100.0 E12.5 F1000.0      ; intro line
G92 E0.0
M221 S{if layer_height<0.075}100{else}95{endif}
```

You do not need to understand G-code to use a 3D printer. To get started, you only need to know that the slicers take a 3D model file and outputs a *.gcode file. That is the file you will hand off to your printer.

Learning basic G-code is useful for printer calibrations and troubleshooting. G-code is a simple language that uses a limited set of individual instructions, so you do not need a software development background to learn it.

Let's dissect this line from the above script:

```
G1 X60.0 E9.0 F1000.0
```

This line primes the extruder by starting some filament flowing just before a print. It tells the printer to move in one direction (G1, "controlled motion") on the X-axis for 60.0mm (X60.0). While driving, the printer should extrude 9.0mm of filament (E9.0) at a feed rate of 1000.0mm/min (F1000.0). The feed rate is how fast the print head moves during this command.

This sort of command is useful for printer calibration. If you marked your filament before running this command, that mark should have moved precisely 9mm. Likewise, the print head should have moved precisely 60mm. If the printer's movements do not match the command, then the printer may require calibration. This is easier for troubleshooting than wasting time and material with test prints.

If you are ever curious to see what your print looks like in code, you can open any gcode file in a text editor. You can also connect a laptop to an FDM printer's USB port and send it commands. Some slicers (more on these below), like Simplify3D, have built-in machine control panels for this purpose.

> **Note**
>
> You will follow the same process when slicing an SLA model, but the exported file will not be a G-Code file. Different SLA printers use different filetypes. For example, the Prusa SL1 uses *.sl1 files. You can still open a .sl1 file like any archive file (e.g., *.zip) and view its contents.

Choosing a Slicer

Your slicer program is an essential tool for 3D printing. There are many different options available, and they are continually changing. Some require the purchase of a software license while others are free and open-source software (FOSS). Some printers even come with custom slicing software the manufacturer created just for that printer. Choosing one to use can be confusing, but it does not have to be. In most cases, using one of the popular free options is a good way forward.

Cura (*https://ultimaker.com/software/ultimaker-cura*) and Slic3r (*https://slic3r.org/*) are two popular FOSS options. You will have no trouble finding online tutorials and configuration information for these programs. Forums and articles frequently reference these slicers, so getting help with a problem is usually easier.

These slicers are open source, so the projects are occasionally *forked* to create new projects. Prusa Research created their slicer, PrusaSlicer (*https://www.prusa3d.com/prusaslicer/*), using the Slic3r codebase and customized it for their printers. If you are using a Prusa printer, PrusaSlicer is an excellent starting point. This slicer has also become a popular option for people who use other printers.

There are also free and closed-source options, like Raise3D's excellent ideaMaker (*https://www.raise3d.com/ideamaker/*), and paid options, like Simplify3D.

The free slicers are likely to serve you very well. While Simplify3D is powerful, the FOSS slicers keep pace with software updates from communities of maintainers. These updates often include exciting new features that won't become available in paid programs for a long time (if ever). FOSS projects are typically more agile and able to experiment with new ideas. The good news is you don't have to pick just one slicer, so try out different options.

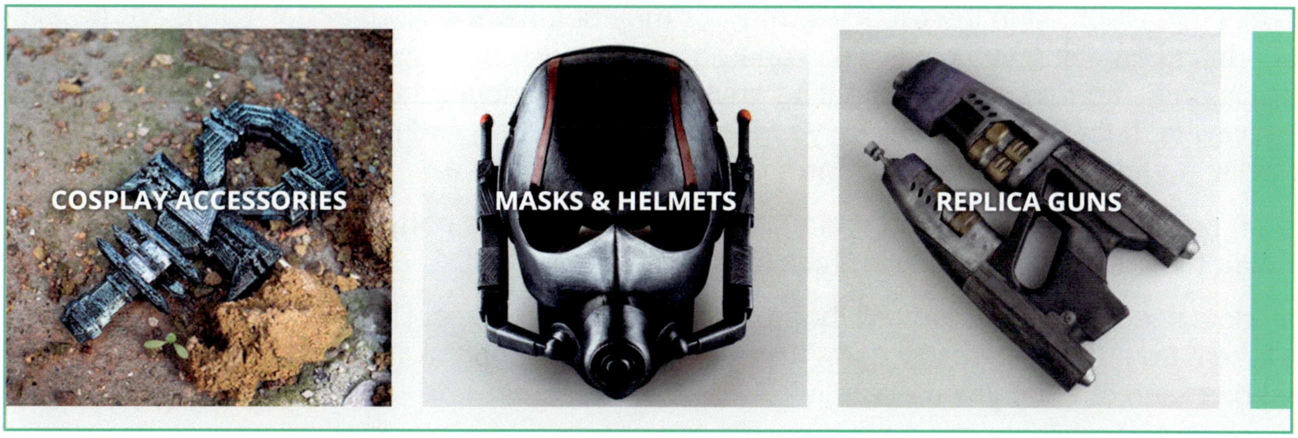

Source: My Mini Factory

You don't need to be a 3D modeler to print something for a costume. There are numerous online websites and communities dedicated to sharing models, such as:

- Prusa Printers, *https://prusaprinters.org/prints*
- Thangs, *https://www.thangs.com/*
- My Mini Factory, *https://www.myminifactory.com/*
- CG Trader, *https://www.cgtrader.com/*
- Pinshape, *https://pinshape.com/*
- Cults, *https://cults3d.com/en*
- Thingiverse, *https://thinigverse.com/*

These sites (and others) offer free models ranging from tabletop game miniatures and fidget spinners to wearable helmets. They have extensive collections of models from games and movies intended for display or cosplay. Some of these sites have opened up shops where amateur and professional modelers sell their designs for modest fees.

The maker community is often generous with the sharing of knowledge and designs. I browse these sites like some people browse social media. Models are usually not vetted prior to upload, so many, many models will not print correctly or work as expected. Check the model's comments to see how well it has worked for others. You can still find examples of incredible works of engineering and design made freely available by their creators.

To highlight just one exceptional design, Sean Charlesworth created a fantastic 1:1 scale design of a Ghostbusters ghost trap. Sean designed this to be as accurate as possible to the hero props used in the movies. He went as far as making it possible to add functional doors, lights, sound, and smoke effects with an Arduino board and some electronic components.

Models like this can take dozens or hundreds of hours to create. Sean published his ghost trap designs for free on Thingiverse (as user *Countspatula*) and opened a shop on Etsy (*https://etsy.com/*) called *Charlseworth Dynamics*, where he sells kits and accessories. This is a popular option for many designers. Some start Patreon (*https://patreon.com/*) campaigns where you can pay for downloads of specific models or pledge a monthly payment to maintain access to repositories of high-quality designs. The cost of entry is $3-10 in many cases. You can support the artists and engineers while embarking on your new project!

Some of the models featured in this book came from Etsy, Patreon, Thingiverse, and other online sources.

Is This Cheating?

Let's address this elephant in the room you may, or may not, have encountered. There are people who insist that printing anything you did not design from scratch is "cheating" or a lesser form of making. Some people go as far as saying 3D printing anything is a shortcut and somehow shameful. Those people are wrong. It feels good to create your own custom pieces and make them real, but it can be equally thrilling to find something like Sean's ghost trap and make it.

These same people probably existed during the sewing machine's early days, insisting those who used a machine over hand stitching were somehow lesser.

When you download a file, you are collaborating with another maker. They provide the blueprints, and then you take it from there by deciding on the scale, material, and finish. That prop is yours, have fun with it and be proud of what you make.

There are also free programs available that enable you to design your very own models. The 3D design process may seem daunting at first, but it is a learnable skill and the best way to get what you want. A later section discusses these programs in more detail.

Regardless of where you get your models, have fun!.

My Version of Sean's Ghost Trap

Link

Sean has a video on *Tested.com*[4] with an in-depth look at his design work. It's an incredible look at how 3D design and printing made this ghost trap possible.

3D File Formats

Most 3D models you find online will download as stereolithography (STL) files. 3D Systems developed this file format in 1987. It works well enough, but it allows for intersecting and non-manifold edges, leading to failed 3D prints. Many slicers try to detect these issues, warn you, and try to repair them, so it's not a significant concern in most cases. Still, it is a common problem with many free models available on sites like Thingiverse. Most sites do not require a creator to prove a model can be printed before they upload it.

> **Note**
>
> The term "non-manifold" refers to the 3D geometry of a model that cannot exist in the real world. One typical example of this is a wall with no thickness. The wall exists in the model but cannot be printed in the physical world. You might also find models where the faces of the model are not attached – they are really just floating next to each other. These issues might cause an unexpected print failure or gap.

STL files lack color, texture, and dimension data. The latter is why you might see your slicer ask if it should import an STL using inches or millimeters. If you select the wrong option, the model will be a different size than intended, and you may not realize it before printing.

If you preview an STL file, your model-viewing software will always load it as some sort of monochromatic mesh (the macOS Preview application seen below).

You may find models shared as 3D object (OBJ) files, the same used by 3D graphic designers. OBJ files can contain color and texture data, so the 3D printing community adopted it for multicolor printers. Even so, some designers still package multicolor models as a collection of separate STL files.

A third rarer filetype is the Additive Manufacturing Format (AMF). These files can contain all of the same data as STL and OBJ files to become the one true 3D printing filetype. It was heralded as "STL 2.0," but has not gained much momentum. You may never encounter it outside of this book.

Viewing the STL File for the Headpiece of the Staff of Ra

Finally, as of 2020, some modelers and slicers have switched to the 3D Manufacturing Format (3MF) filetype developed by the 3MF Consortium (*https://3mf.io/*). Like AMF, this file format aims to replace all of the other file types and make models much easier to share. Unlike AMF, the 3MF filetype has been gaining momentum.

The 3MF Consortium has several large corporations (e.g., Microsoft) and 3D printer manufacturers behind it. For example, the slicer released by Prusa Research, PrusaSlicer, saves projects as 3MF files by default at the time of this writing[5].

The technical aspects of these filetypes are beyond the scope of this book. Slicers can open and process all of these file types. Hence, the file's technical details are often inconsequential when you just want to print a model from the internet.

That said, there is one reason to be excited about 3MF files. These files can contain printer and support material settings, so a creater can share their model and the printer settings they found work best for that model. This reduces setup time and removes guesswork while still allowing you to make edits. This is an excellent feature for sharing models with beginners and maker communities.

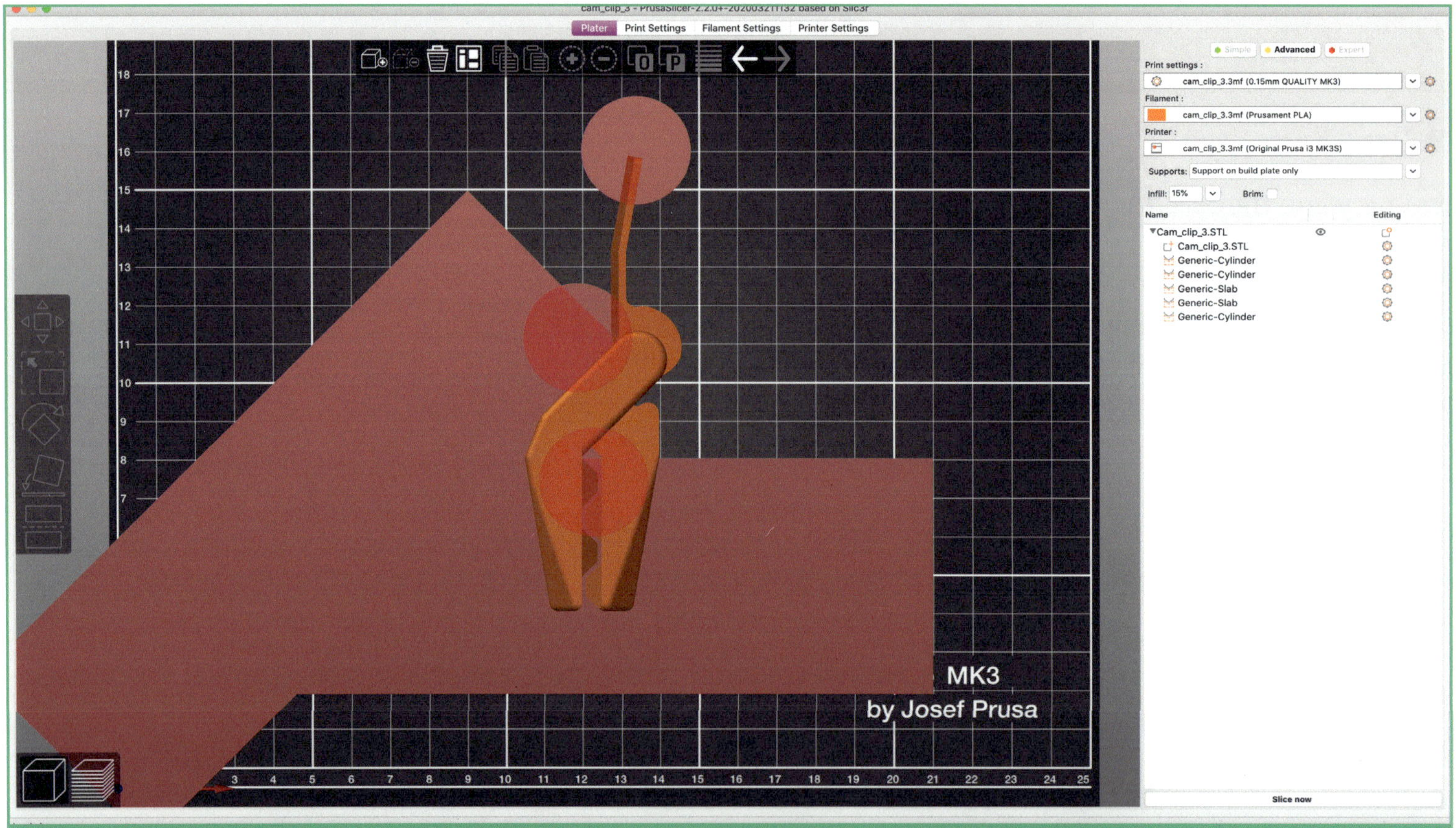

Imported Settings from 3MF

This example shows PrusaSlicer opening a 3MF file and loading all configuration data needed to slice and print this awesome bag clip model by Prusa user *Andrei*. The orange pieces are the parts to be printed. The light red blocks are support enforcers that tell the slicer were to add support material. The 3MF file provided all of this, including the position of the parts and other settings (e.g., infill percentage, layer height).

Note

Check-out the Bag Clip model by *Andrei* on Prusa Printers[6].

Useful Accessories

Printer manufacturers frequently ship a small collection of consumable goods and tools with new printers. You can still improve the time you spend with your printer by upgrading a few standard accessories and stocking up on consumables.

Consumables

First, there are a few consumable items you should have in stock in your shop. The most ubiquitous consumable for SLA and FDM printers is Isopropyl Alcohol (IPA). This stuff washes away resin, helps remove oils and grease from build plates, and cleans printer parts during maintenance. Keep a squirt bottle of IPA nearby; you'll reach for it every time you use your printer.

IPA percentages (how much of the solution is alcohol vs. water) can vary by region. A 70% IPA works well for cleaning resin prints and wiping down print beds, but might be more scarce. The 70:30 mix is gentler on skin, so demand is high for use as a sanitizer for hands and surfaces. You might have an easier time purchasing a jug of 90, 91, or 99% and diluting it yourself.

91% IPA as Found in the United States

For SLA printers, the FEP film at the bottom of your resin tank is also a consumable. It will wear out over time and need to be replaced. Buy some spares before that happens, just in case you accidentally damage or puncture the film.

Heating Block with Broken Heater Cartridge

While not a consumable, per se, consider buying an assortment of critical spare parts for repairs. One of the easiest and most common ways people damage an FDM printer is rough handling of the delicate wires that connect to the heat block's thermistor and heating cylinder.

These wires are easy to break and are often some of the first things to get get covered in molten plastic during major print failures (the dreaded extruder blob). Pulling that plastic away from the connections or cleaning the extruder a little too roughly can snap these wires.

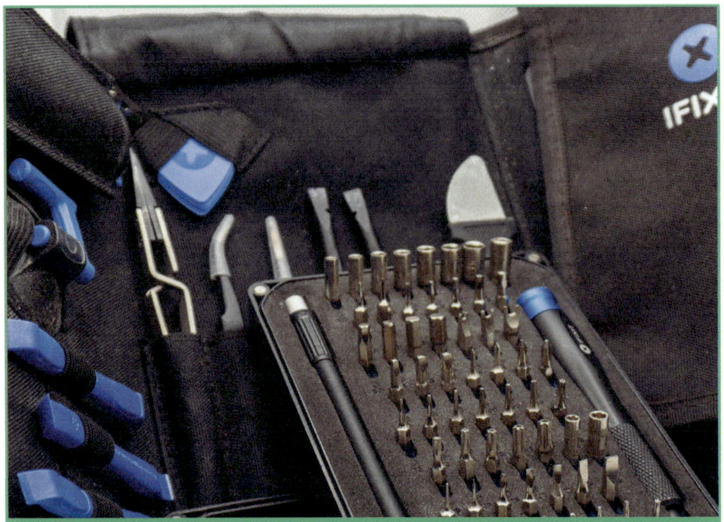

Contents of the iFixIt Pro Kit

Your printer will come with an assortment of allen keys for assembly and maintenance, but allen keys are often tricky to use in the tight spaces of some printers. Consider purchasing a set of hex screwdrivers. Drivers from Wiha and Bondhus have excellent reputations and will serve you well. The electronics repair kits from iFixIt include a great collection of bits and other handy tools. Even if you don't plan on fixing electronics, the other tools in their Pro kit have alternate uses for 3D printing. The pry bars and spudger work well as applicators for body filler, and tweezers are always helpful when working with small screws, little nozzles, and tiny bits of plastic.

It is also helpful to keep a few other tools near your printer:

- 7mm socket wrench for swapping your nozzles
- Sharp flush cutters and pliers for gripping and pulling FDM support material
- A scraper with changeable plastic blades for safely and easily removing material from print surfaces

Brush, Drivers, Pliers, and Plastic Scraper

To round out your printer toolbox, pick-up a brass bristle brush. These brushes work magic on heat blocks, cleaning-up gunk and bits of melted plastic. You can find them marketed as 3D printing accessories with high price tags online, but check-out the welding section of your local hardware store. You can find brushes for as little as $2.

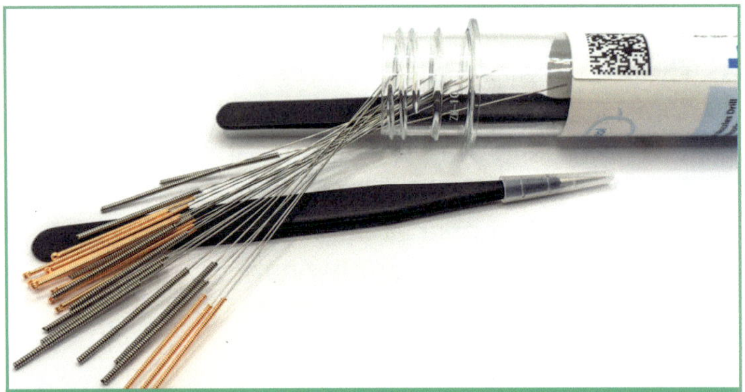

Acupuncture Needle and Tweezers Kit

For smaller clean-up jobs, get some tweezers and acupuncture needles. These work well for grabbing stray bits of filament and clearing clogged nozzles. The needles are flexible and 0.4mm thick. You can gently insert them into a heated nozzle and clear minor clogs. You can find kits with a collection of needles and tweezers for around $10.

Heat Gun

The heat gun deserves its own section. One of these can quickly warm-up a print to help you manipulate it for repairs and finishing work.

Not a Hair Dryer

A few quick passes with a heat gun will clean-up any “stringing” (thin, web-like bits of filament) or little “cobwebs” of filament on a fresh FDM print. The strings melt and disappear with little to no trace.

FDM prints can also suffer from crazing, a collection of micro-cracks, on surfaces that printed facing support material. Crazing changes how the surface reflects light and has an adverse effect on the model’s color. If you’d prefer not to sand and paint the part, a few quick passes with a heat gun can restore the plastic’s original color.

This print of a jack-o-lantern required some support material in the eys and mouth. It completed with some light PLA “cobwebs” and had stubborn support material stuck in its teeth. A few passes from the heat gun fixed both issues.

Before the Heat Gun

After the Heat Gun

Likewise, you can carefully apply more heat to fix a warped part. If you have a plastic or resin part bent or twisted, the heat can coerce it to return to its original shape.

You can also do this to create intentional changes to the model’s shape to modify a fit or add dents and other “battle damage.” The ***Finishing 3D Prints*** chapter explores this idea.

Before you start heating anything, place your part on a surface that can handle the heat. You can easily start a fire or scorch something. Even if your surface can resist the heat, make sure it won’t transfer the heat to something below it or nearby that will melt or ignite!

The typical stock nozzle for an FDM printer is a brass nozzle with a 0.4mm chamber. Brass conducts heat very efficiently, so these nozzles heat up quickly and stay hot. Brass is a soft metal, however. The nozzle will wear down quickly if you print with abrasive filaments. When a nozzle begins to wear out, the chamber's diameter and the shape starts to change. A worn nozzle hurts the quality of your prints.

Hardened steel nozzles are available for abrasive filaments. They work as well as brass nozzles, but steel conducts heat less efficiently than brass. You may need to print with temperatures 5-10°C hotter than usual when using these nozzles.

There are other types of nozzles, like copper-plated and tungsten carbide, but brass will serve you well for most of your 3D printing needs, and steel will cover the abrasive filaments.

Of course, 3D printing technology is continuously changing. Many hobbyists enjoy trying new types of nozzles and extruder systems. If you have a passion for working with hardware, you may find yourself with a growing collection of nozzles and extruder upgrades.

Note

As a rule, layer height should never exceed 80% of your nozzle's diameter. A 0.4mm nozzle cannot print layers taller than 0.3mm.

Nozzle Diameters

Nozzle diameter has a significant impact on the time a print takes to complete and the detail and strength of your 3D printed parts. Josef Prusa, the founder of Prusa Research, once aptly compared printer nozzles to camera lenses. Like swapping lenses on a camera for different types of photography, never be shy about switching your nozzles to set up your printer for better prints.

Layer height impacts your print's vertical (Z-axis) resolution, while nozzle diameter affects the horizontal (X and Y axes) resolution. That means the nozzle affects the level of detail you get on the plane *parallel* to your print bed, so if the top of your model has tiny details or fine text, a smaller nozzle will capture those details much better than a stock 0.4mm nozzle.

If you plan to print large pieces that don't have any delicate details, swap in a nozzle that is 0.6mm or bigger. This simple swap can save you hours of print time. The trade-off is thicker layers and some loss of detail, but if you are printing pieces with a lot of mass and few small details (think a big helmet or armor piece), a bigger nozzle will save you a sizable chunk of time. The fatter layer lines even add strength to your part.

Let's say you wanted to print the previously mentioned model of a Banzai Bill from *Mario Bros.* at a 200% scale. Using PrusaSlicer's default settings for 0.2mm layers with a 0.4mm nozzle, printing the main body would take 20h 14m. Swapping to 0.3mm layers with a 0.6mm nozzle slashes the print time in half for 10h 7m.

Sliced Info	
Used Filament (m)	73.57
Used Filament (mm³)	176953.68
Used Filament (g)	219.42
Cost	5.48
Estimated printing time:	
- normal mode	20h 14m
- stealth mode	20h 22m

Sliced with 0.2mm Layers and 0.4mm Nozzle

Sliced Info	
Used Filament (m)	87.55
Used Filament (mm³)	210587.88
Used Filament (g)	261.13
Cost	6.53
Estimated printing time:	
- normal mode	10h 7m
- stealth mode	10h 12m

Sliced with 0.3mm Layers and 0.6mm Nozzle

A 0.25mm nozzle gets better detail from 0.1mm or 0.05mm layers. This is great if you are printing text or a piece of jewelry for a costume. These prints will take much longer to complete, however. These tiny nozzles should only be used for small parts.

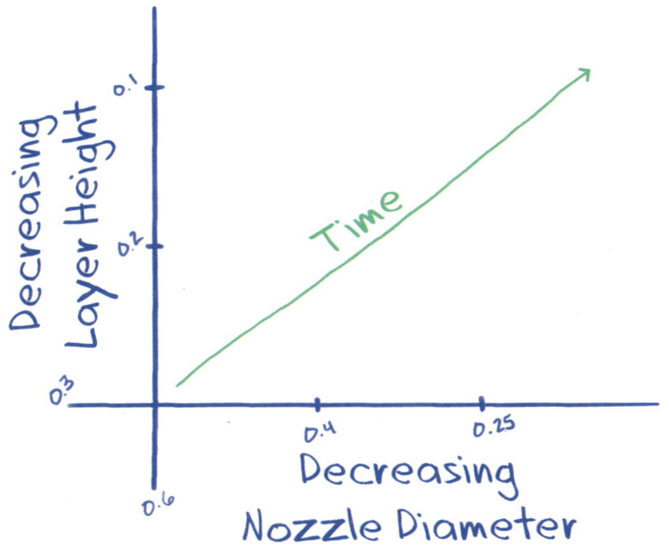

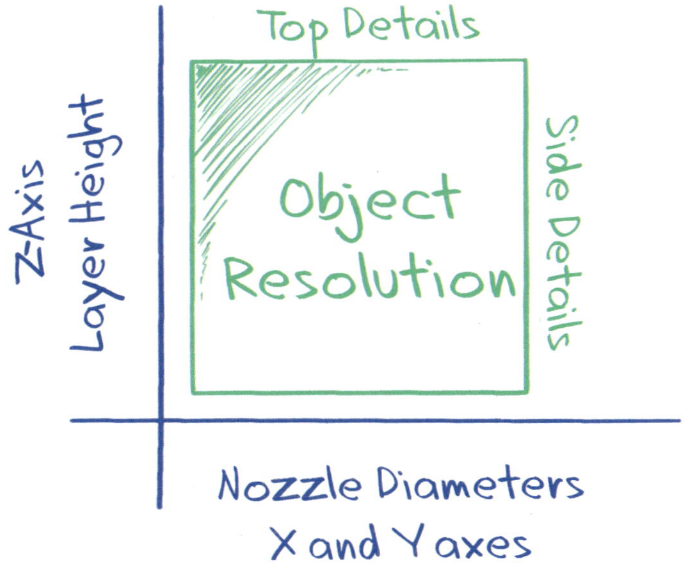

Basically, decreasing the nozzle diameter or layer height increases print time. That should not dissuade you from making adjustments. When you are preparing to print a model, review the locations of the details. The layer height impacts the details on the vertical faces, and the nozzle diameter controls the resolution of the horizontal (top and bottom) layers.

When I modeled this Nuka Cola bottle cap from the video game *Fallout*, I wanted a version with the text for multi-color prints. The text doesn't look too small in this rendering (the picture is roughly 1:1 scale with the printed model–32.10mm outer diameter), but some of the thinner lines of text are 0.3-0.39mm wide, tapering to ~0.28mm.

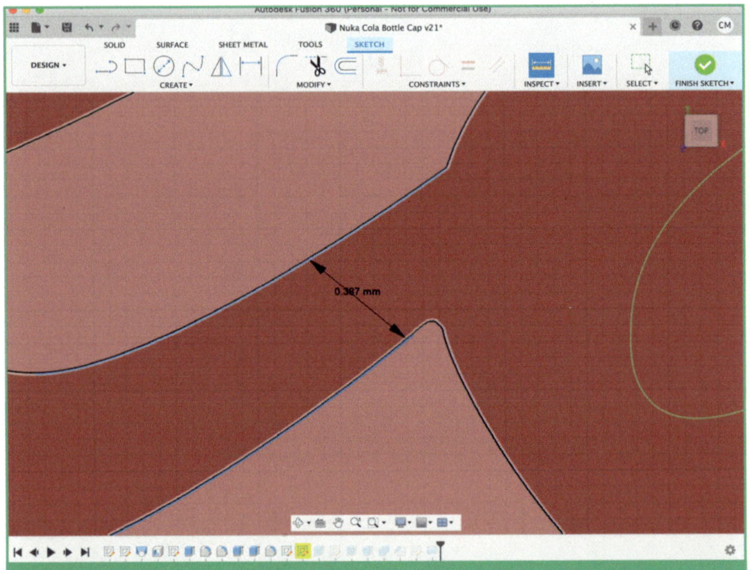

Measuring the Width of the Text

The text is on top of the model, so the nozzle diameter determines if those lines can be printed. A 0.4mm nozzle just can't print those lines. Here is the cap sliced using a 0.4mm and then a 0.25mm nozzle.

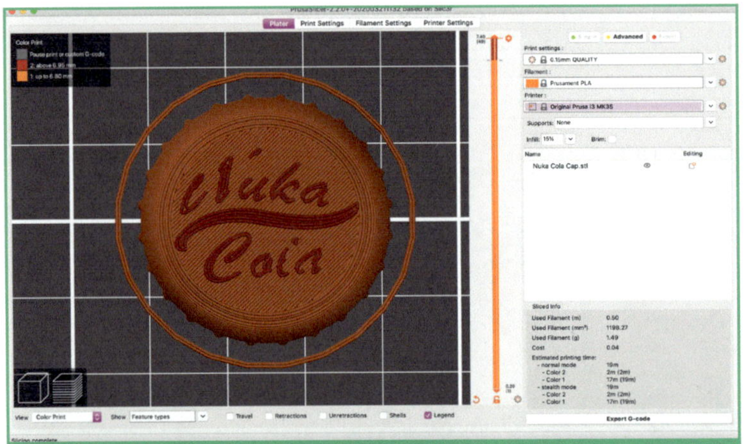

Sliced with a 0.4mm Nozzle

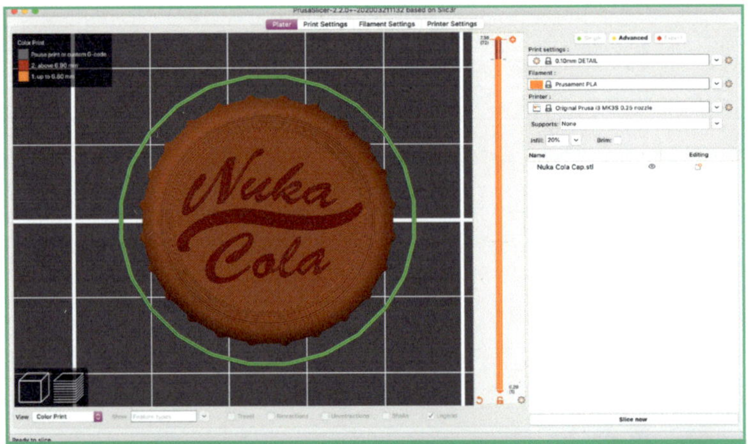

Sliced with a 0.25mm Nozzle

For layer height, print at 0.1mm or above for the best ratio of print time to print quality. You will see minimal improvement with 0.07mm and 0.05mm layer heights. Prints take twice as long to finish for imperceptible gains. Remember, the final part is most likely going to be sanded and painted, so don't torture yourself by extending print times unnecessarily.

I used a 0.25mm nozzle and 0.1mm layer height to print these multi-color Nuka Cola bottle caps.

DL-44 Heavy Blaster

Star Wars

Model by PortedReality[7]

PLA, wood composite PLA, and cold-cast urethane resin

Finished with Tamiya acrylic paints, Testors metallizer lacquers, oil paints, and graphite powder

Printing Materials & Storage

There are a wide variety of printing materials available from numerous brands. This section will focus on the basic material types that are most commonly used for props and cosplay, how to choose the right material for your project, and how to store your materials.

Materials covered in this section:

- Polyactic Acid (PLA) Filament
- Acrylonitrile Butadiene Styrene (ABS) Filament
- Glycol-Modified Polyethylene Terephthalate (PETG) Filament
- Flexible Filaments
- Composite Filaments
- Nylon Filament
- Photopolymer Resin

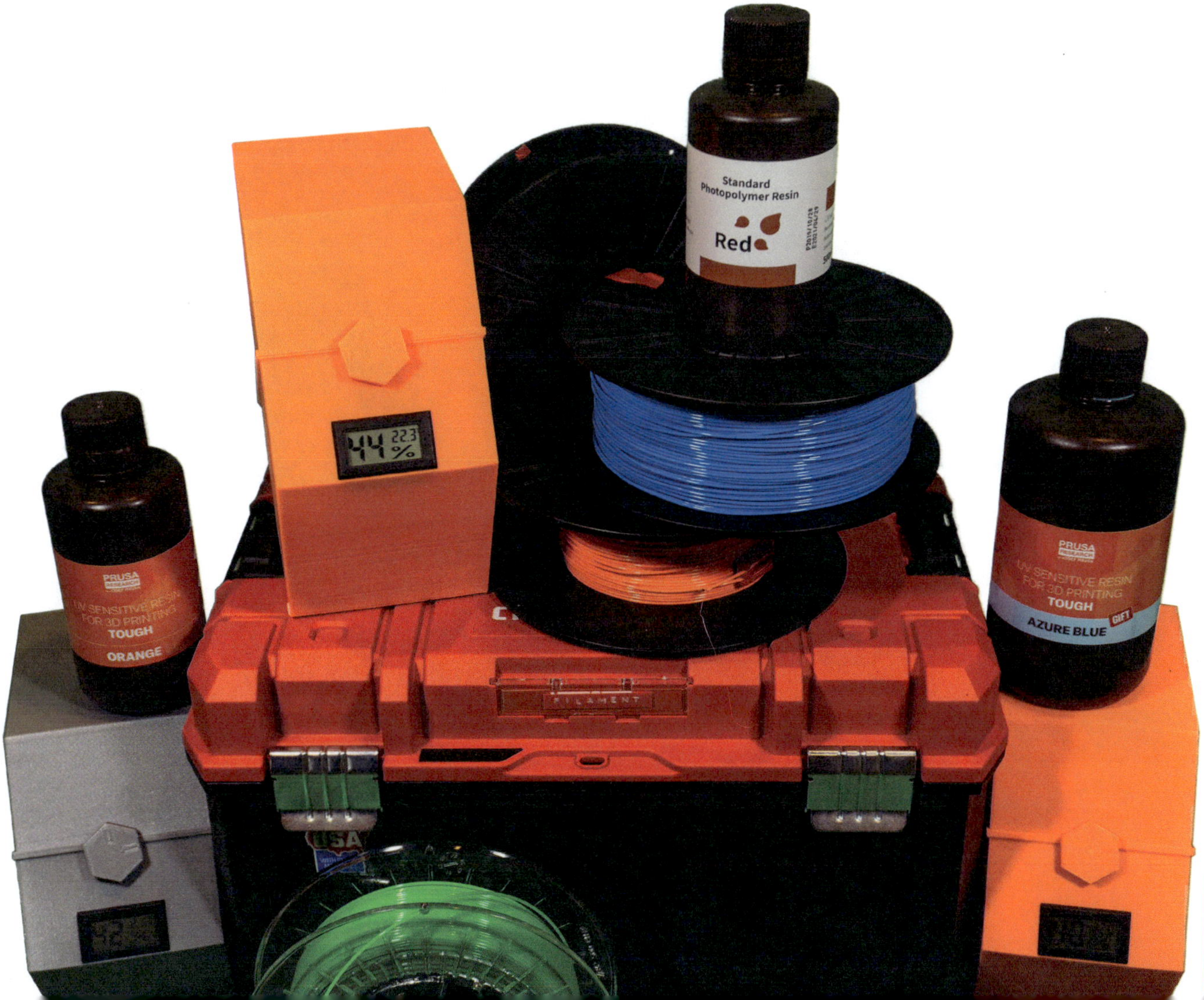

Polylactic Acid (PLA) Filament

The number one thermoplastic used with FDM printers is polylactic acid (PLA) filament. PLA is the most beginner-friendly filament and it is biodegradable (made from sugarcane and corn starch). It works well for all kinds of models. Manufacturers use PLA to make items like tea bags and disposable cups.

PLA filament comes in a vast array of colors. It can even be translucent, contain glitter for sparkles, or have a metallic appearance. Suppose you want a part to be a specific color without any paint. In that case, there is probably a spool of PLA available in your color.

Most PLA has a glossy finish, but you can find matte filament. The glossy finish of regular PLA reflects light that can accentuate layer lines. A matte finish naturally hides layer lines for a cleaner look, so it is an excellent choice for prints that you don't want to paint. .

I printed this Banzai Bill from *Super Mario World* in matte "coal black," "paper white," and "jujube red" from SpiderMaker.

Banzai Bill Printed in Matte PLA

Printing PLA

PLA is a very beginner-friendly material. It has a low melting point of 150-160°C. Manufacturers provide recommended extruder temperature ranges for their PLA that are usually around 180-220°C. PLA also has a low thermal expansion, which means it does not shrink much when it cools.

PLA begins to melt at 160°C but needs to be heated to at least 180°C to flow through a nozzle for printing. 190°C is usually the lowest temperature recommendation you are likely to see on a spool. Some of the most popular brands print at 210-220°C.

In addition to its low melting point, PLA also does not give off any harmful fumes or bad smells when it melts. Usually, if you smell it, you might get a hint of something sweet in the air.

Printing & Prop Considerations for PLA

PLA's disadvantages are something to consider when it comes to cosplay and prop making. PLA is sturdy and rigid, which makes it difficult to sand. It is easier to get a smooth surface with some other materials. The rigidity also means PLA snaps or cracks much sooner than some other plastics if bent or dropped.

PLA's low heat resistance is a potential concern. PLA starts to melt at 160°C but begins to deform in temperatures above just 60°C. Your prints may find themselves exposed to these temperatures when you least expect it, like if you left a prop in your car on a hot summer day. You may return to find your prop or costume piece deformed by the heat.

Note

PLA's low heat resistance also affects sanding PLA models. Sanding generates heat. A mix of heat, pressure, and small print issues may lead to problems like layer separation (delamination) with the outer walls (shells). You have to be careful and work slower than you might on another material. It is better to wet sand (use some soapy water) PLA and work slowly.

Cost of PLA

PLA generally costs around $20 for a 1kg (~2.2lb) spool. PLA is a cost-effective filament for rapid prototyping and learning. Some spools are priced closer to $35 (or higher), but these are often specialty types of PLA, like exotic colors or composites.

What About PLA+ Filament?

Some manufacturers now offer a so-called PLA+ filament. PLA+ is PLA with additional materials mixed in, like other plastics. That's why you might see it called "PLA with additives." The added ingredients are intended to address some of PLA's disadvantages, like deforming too easily above 60°C, so PLA+ tends to have a higher melting temperature. It is also supposed to be have increased durability with more vibrant colors. It has been called "the missing link" between PLA and the tougher filaments (more on those later).

The trouble is each manufacturer has a unique PLA+ formula, so it is difficult to speak generally about any advantages of PLA+ has over regular PLA. In theory, PLA+ offers a filament option that remains beginner-friendly while gaining some durability.

Acrylonitrile Butadiene Styrene (ABS) Filament

Many prop makers eventually transition from PLA to acrylonitrile butadiene styrene (ABS). LEGO uses ABS to make their bricks. You probably have many other items in your house that have ABS parts. This material is strong and has the flexibility PLA lacks. It contains high-wear synthetic rubber, is impact-resistant, and resists heat much better than PLA (100°C vs. 60°C).

ABS is similar to PLA when it comes to filament selection. ABS is available in a range of colors, but options tend to be more limited than PLA. Color selection is usually less of a concern when buying ABS for cosplay or props because you will probably sand and paint the finished model.

ABS Prints Look Great Once You Get the Hang of It

Printing ABS

ABS is an amorphous polymer, which means it does not have one melting point. For 3D printing, this means there is a broader print temperature range of about 220°C to 260°C. Each manufacturer offers a temperature recommendation for its filament, but you'll need to experiment with any new filament you get to find what works best for your printer and environment.

When ABS melts, it can smell bad and create fumes. It is best to print ABS in a well-ventilated room. You don't need anything special for ventilation; an open window works just fine. Some people report ABS fumes give them headaches, even dizziness. If you find you are in this group, take care to place your printer somewhere away from you with good air flow.

Printing & Prop Considerations for ABS

ABS has a reputation for being difficult for beginners to use because it is so particular about temperature. A fan creating a light breeze (including the printer's cooling fans) can be enough to cause the ABS to cool too quickly or not reach the necessary temperature to bond to the previous layer. The printer's fans may need to be turned off or lowered to 10-30% power for ABS prints. This is tricky because a lack of airflow can also prevent proper cooling on small sections of a model and cause deformation. You may need to experiment with settings for your printer.

ABS also has a high thermal expansion–i.e., it shrinks as it cools. This is another reason why rapid cooling leads to significant problems or outright print failures. ABS parts won't shrink so much that they will be noticeably smaller to the naked eye, but this shrinkage will cause parts to pull against the print bed and potentially lift up at the edges. This loss of contact with the print bed is the leading cause of ABS print failures. This effect is referred to as "warping" and manifests as flat edges of a model curling up and causing the shape of the following layers to be warped to varying degrees.

To combat cooling and warping issues with ABS, place the printer inside of an enclosure. Some people use simple cardboard boxes while others assemble prettier enclosures from IKEA furniture parts. The enclosure aims to protect the print from drafts and keep the air temperature around the printer warm (~35-40°C).

ABS has a few features that make dealing with any temperature issues worthwhile. For one thing, acetone dissolves ABS plastics. You can smooth ABS prints with a process called "vapor smoothing," which involves suspending an ABS print above a bowl of acetone and heating the acetone. The acetone quickly turns into vapor and begins dissolving the print's surface. The vapors are just enough to eat away at the surface and remove layer lines.

Note

ABS shrinkage is typically around 7%. You won't notice this with your eyes, but you will see it if you expect the complete print to be a specific size. Shrinkage is why ABS is not the right choice of filament for printer calibrations. You may also need to account for the shrinkage in your designs if you combine an ABS part with other materials.

Vapor smoothing is neat, but ABS also sands more easily than PLA. Sanding is usually much more convenient and definitely less messy.

Acetone can also help you fill gaps and bond separate pieces together. You can create a few different solutions by mixing failed prints and leftover filament with acetone. These solutions are typically called ABS juice, glue, and slurry. The ***Gluing Printed Parts*** section covers these mixtures in-depth.

Strength, ease of sanding, and ABS glue make ABS an appealing material once you are comfortable with 3D printing.

Warning!

Acetone is a serious chemical. It is highly flammable, gives off harmful fumes, and must be handled with care! If 3D printing is something you plan to do with young children, keep any acetone and ABS mixtures tucked away in a safe place.

Cost of ABS

ABS usually costs around the same as PLA, about $20+ for a 1kg spool. It has a steeper learning curve, leading to more failed prints and waste material when you first get started. You can turn waste into ABS juice, glue, or slurry, so early failures do not impact affordability very much.

What About Acrylonitrile Styrene Acrylate (ASA) Filament?

ASA is sort of a new generation of ABS. All of the above information applies to ASA. What sets ASA apart from its predecessor is its resistance to sunlight and the elements. If left in direct sunlight, ABS will fade and yellow. ASA stands up to weather and sunlight better and retains its color. This is one reason automotive manufacturers use ASA for car parts that will be in direct sunlight.

If you have ever set a LEGO model or similar plastic item on a windowsill, you have probably witnessed how the ABS plastic can yellow or lose its color. ASA is much more resilient in that area.

From a cosplay and prop perspective, there is no reason to switch to using ASA over ABS. Both perform just as well for most cases but consider using ASA if you are making something that will be displayed outdoors or in front of a window.

Glycol-Modified Polyethylene Terephthalate (PETG) Filament

Polyethylene Terephthalate (PET) is one of the most used plastics today. Much of its production goes towards making bottles and synthetic fibers for clothing (polyester). PET has an incredible strength to weight ratio and is virtually shatterproof. Additionally, PET is ductile (good flexibility without losing strength) and resistant to water and chemicals.

For 3D printing, manufacturers modify PET with glycol (the G in PETG) to make it easier to print. As a filament, PETG comes in a decent selection of colors, including transluscent options that set PETG apart from other filaments. Printing with transparent PETG filament produces models with some fantastic light refraction as layer height and print nozzle diameters are changed.

Transluscent PETG Vase

Pure PET filament does exist, but it is much more difficult to print. If you need to print PET for a functional part for your printer, a household item, or a costume, PETG is probably your best option. As you will see, PET is more of a niche option for cosplay and props, but useful for specialty parts (e.g., functioning gears or transluscent parts).

Printing PETG

PETG requires temperatures between 220°-250°C. When it comes to adhesion, PETG loves to stick to itself and surfaces. Each new layer bonds firmly to the previous layer, creating durable prints that are waterproof and look great. This is a blessing and a curse; PETG can adhere too well to the print bed, leading to damaging some print surfaces during print removal. To avoid this, you should use an "interference layer" (separator) between the print bed and the filament.

Check your printer's recommendations for PETG as print beds differ between printers. Common interferencer layers are Windex glass cleaner and water soluable glue sticks (e.g., Elmer's Disappearing Purple School Glue).

PETG has a low thermal expansion, so warping is less of a concern. You can print PETG without an enclosure, like PLA.

Printing & Prop Considerations for PETG

PETG may sound like the perfect universal filament. Many people do pick PETG as their "default" filament. Yet, there are several factors to consider before using PETG for a prop.

The first issue is that PETG is hygroscopic. It wants to absorb water. You must keep PETG filament dry (low humidity) while in storage and during printing. You need to keep all types of filament dry, but PETG will more readily absorb moisture from the air. The ***Filament Storage*** section discusses proper storage and how to identify and dry filament that has absorbed too much moisture.

Even when you store it properly, PETG will absorb what little moisture is in the air around it. It is common to see people recommend drying PETG for a couple of hours before printing with it.

Drying the filament before each use does tend to improve the results, but also costs additional time and energy.

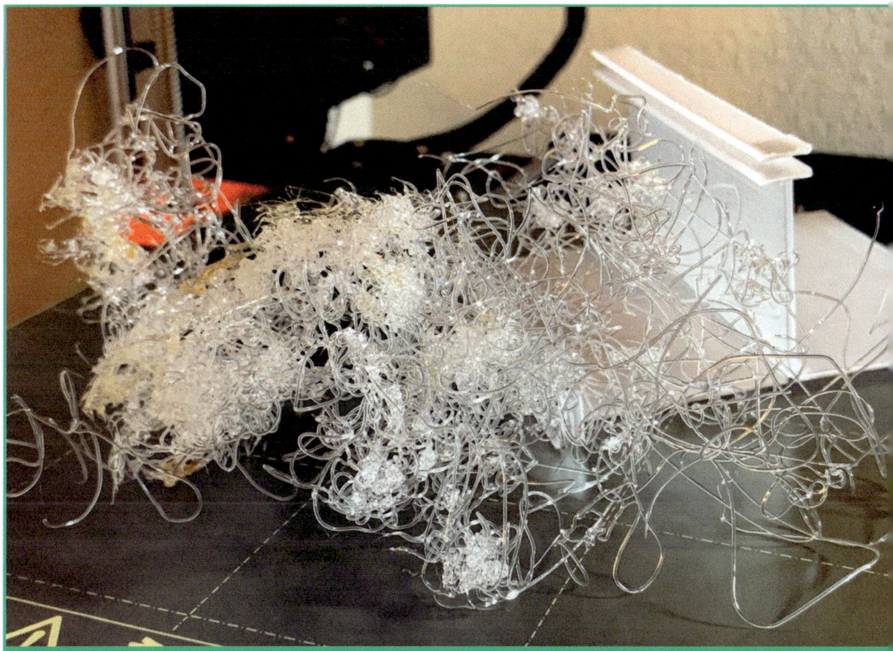

The PETG Blob Won This Round

For cosplay and props, PETG's water and chemical resistance can work against you. Even if a model is printed in an appropriate color for the final piece, you may still want to paint parts of it or weather it, and PETG is challenging to paint. You can paint it, but the paint may not have a strong bond.

PETG also scratches easily. This makes it as easy (maybe easier) to sand than ABS, but it also means you might get unwanted scratches. A finished PETG piece will likely scratch and lose paint with handling. Keep this in mind when using PETG to make anything you want to frequently handle, wear as part of a costume, or pack for a convention.

You can take steps to improve paint adhesion and scratch resistance. One option is applying a layer of Smooth-On XTC-3D, a self-leveling epoxy intended for finishing 3D prints. This epoxy creates a smooth, glossy surface that you can sand and paint.

Finally, PETG likes to stick to the brass nozzles that come stock on most 3D printers. The brass nozzles tend to pick up strings of PETG that grow into blobs of filament as more strings become stuck. Eventually, the lump will drop into your print or catch a wall, creating a rough surface or causing a more catastrophic issue. You can purchase nickel-plated nozzles that don't pick-up strings of PETG like brass. Usually, you can fix this by tinkering with your slicer settings to dial-in the perfect first layer settings.

Cost of PETG

PETG filament costs about the same as PLA, ABS, and ASA–around $20-30 for a 1kg spool. Consider buying some PETG if a part needs to be waterproof or transluscent.

Note

PET may be shatterproof when used for an injection molded bottle, but your PETG 3D prints are not indestructible. PETG prints are tough and durable, but you can easily snap them along layer lines or crack them with an overtightened screw.

Flexible Filament

Thermoplastic elastomers (TPE) are a category of flexible thermoplastics that opens up some fun 3D printing possibilities. Thermoplastic polyurethane (TPU) is the most common TPE in 3D printing. This is the same material used for many phone cases. It's rugged but flexible enough, you can use it to print items like custom reusable zip ties.

There is a much more limited color range for flex filament. You can find transluscent, primary colors (e.g., reds, greens, and blues), and shades of black and white.

You may see TPE filament generically referred to as "NinjaFlex" when browsing 3D printing forums. There are many manufacturers of TPE filament, but NinjaTek became well known as a source of this filament early on with their line of NinjaFlex filament. It's akin to how people tend to refer to adhesive bandages as Band-Aids no matter who manufactured them.

Flex is useful for things that will see heavy wear and tear, like cases and toys. Think of items that might need to bend or take an impact. You can also print things like custom wheels for LEGO models or rubber-like grips for a prop blaster.

Look at how these pliers crush a 3DBenchy boat. Benchy snapped right back into shape.

3D Printed Reusable Zip Ties

Benchy Was Just Fine

Printing Flex

TPE has a melting point between PLA and ABS. The range is about 225° to 245°C.

Some machines may not be capable of printing with TPE filament very well, depending on the feeding system. The flexible filament might bend or twist when the printer pushes the filament forward, leading to the filament binding or twisting. TPU is more rigid than TPE, so TPU filaments tend to be easier to print.

Printing & Prop Considerations for Flex

When evaluating TPE filament options, pay attention to the Durometer Shore A Hardness rating. This is a number followed by an *A*, like *95A*. The Shore A Hardness scale is used to measure the hardness of flexible materials. Many the TPE filaments have a Shore A Hardness of 95A. This filament will feel soft and flexible compared with other filaments but still falls under "Hard" on the Durometer scale.

Flex filament flops around like a cooked noodle. You'll never confuse it with another filament.

A 95A hardness is similar to a rubber wheel on a shopping cart. If the object is thin, a 95A material will still twist and bend. Some filament is rated as 85A and slightly softer. This filament creates parts that feel more like the sole of a shoe.

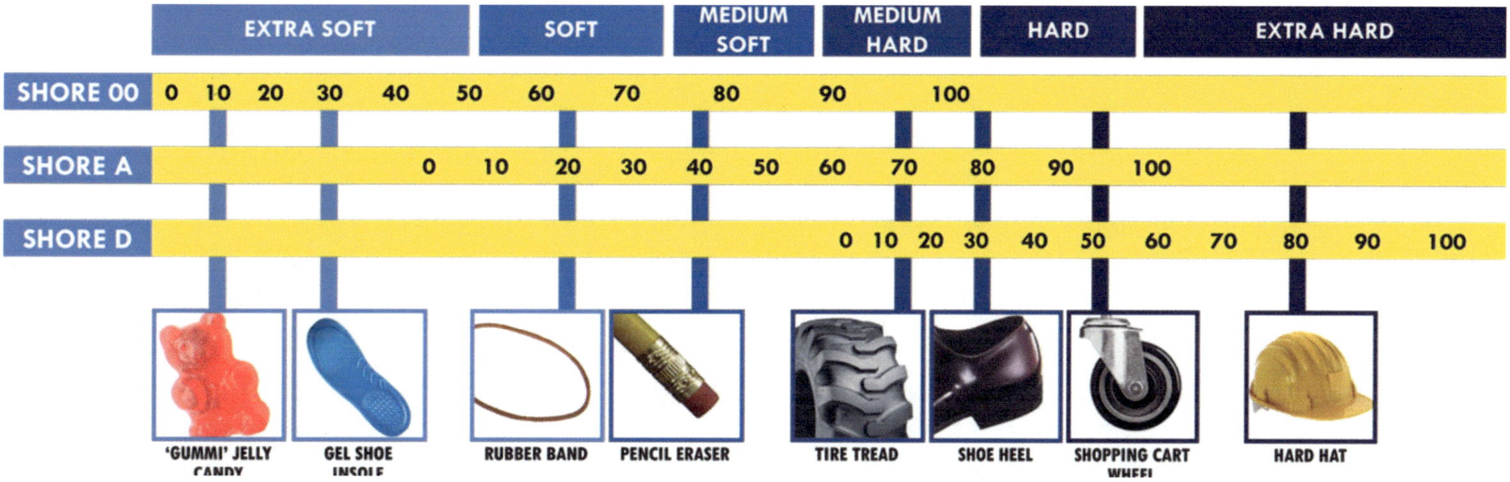

Source: Smooth-On Shore Hardness Scales, *https://www.smooth-on.com/page/durometer-shore-hardness-scale/*

Another consideration is color and finish. You cannot sand and paint flexibile prints like other prints. Similar to PETG, the material is chemically resistant, so paint will not adhere well to the surface. Further, many paints will crack and flake when flexed. There are flexible paints that work well on pliable materials like EVA foam, but even they may crack if repeatedly bent and pinched. Expect to print in one of the primary colors and skip sanding and painting.

It's not impossible to paint flexible prints. The paint finish may just not last a long time if you are frequently crushing, squishing, or bending the printed pieces.

Cost of Flex

The price of flex filament has a wide range. NinjaTek sells its 1kg spools of NinjaFlex filament (TPE) for ~$85. Other brands sell spools of TPU filament for ~$30 per 1kg. Price tends to be affected by the Shore A hardness. Some of the more expensive filament (NinjaFlex) is 85A and excellent for wearables and items you want to be extra floppy or squishy. The less expensive filament is usually 95A or 98A. Think about that Shore A Hardness number and what you want to make with the filament. It may take sense to pay more to have the correct material.

Retailers and manufacturers often sell the pricier flex filaments in smaller 500g spools. You get less material for the same cost as other brands, but flexible filament has niche use cases. One 500g spool will likely give you plenty of filament to experiment with and determine what will work best for you.

Composite Filaments

Composite filaments are usually PLA, ABS, or ASA mixed with another material to create something new and exciting. Many fun composite materials range from glow in the dark filament to filament that conducts electricity.

One of the most basic composite filaments is a filament infused with wood fiber (bamboo, cork, cedar, and other types). This filament sands and accepts wood stain just like wood, so you can create some fantastic creations that have the appearance of being carved from wood. Creative use of orientation and top/bottom fill patterns in your slicer software can add a pseudo wood grain effect to the final piece.

I used a wood filament to create grips for this replica of Han Solo's DL-44 blaster replica from *Star Wars*. Usually, slicers default to to using a rectilinear pattern (straight lines) for the top and bottom layers of a print. I used a circular pattern for the top layers to give the grips a more natural appearance. Once stained and lightly weathered with oil paints (to provide some dirt and a nice oily sheen), they gave the impression of wood.

Like the wood filament, there are filaments mixed with metallic powders made from copper, brass, and other metals. Parts printed from these filaments are weightier and can be polished like metal.

There are other more exotic composite variants, like ferromagnetic filament and conductive filament. You can use these filaments in low-voltage circuits. The possibilities for this stuff in a prop or costume are exciting.

Printing Composites

For the most part, composites print like their base filament (e.g., PLA). You may need to increase printing temperatures for some of them. These filaments even look different when printed at different temperatures, so you will want to play with them. This goes double for wood. After all, it's made from wood so it will darken when exposed to higher temperatures.

Some experimentation may be required to get whatever effect you desire using these new materials. For example, glow in the dark filament often looks its best when printed with minimal infill and thick walls. It needs the mass to absorb light.

Also, these filaments tend to print best with larger nozzles. The standard 0.4mm nozzle will clog more easily with composite filaments. It is best to use a nozzle that is 0.5mm or wider, if you have the option.

Printing & Prop Considerations for Composites

The majority of composite filaments tend to be abrasive, especially those involving powdered metals. Printing with these filaments will significantly increase the wear and tear on a standard brass nozzle. If you want to use metal or cartbon fiber filament, consider buying a hardened stainless-steel nozzle.

Cost of Composites

The cost of composite filament varies by type and are usually more expensive than their base filament. Glow in the dark and wood-filled filaments tend to be less expensive (closer to the cost of a premium spool of PLA or ABS). Metallic composites are usually much more costly, with prices north of $50 per spool.

Blaster Grips Printed in Wood PLA and Colored with a Mahogany Wood Stain

Photopolymer Resin

Photopolymer translates to "light-activated" polymer. This is the type of material used by SLA printers. An SLA printer cures the resin using UV light to produce a highly detailed resin part.

Printing Resin

This part is straightforward compared to FDM filaments. You must match the type of resin to the printer. Different resins cure at different UV light wavelengths. The manufacturer marks the kind of resin with nanometers (nm) measurements on the bottle. The standard SLA printing wavelengths are 365nm and 405nm. The measurement on the bottle needs to match the printer. Follow the instructions for your printer to pick the right resin.

Darth Vader Bust Printed in Resin

Resin gives off a strong smell and fumes, so keep the area well ventilated when using the printer and during the printing process. Also, use nitrile gloves when handling the resin. Resin is a skin irritant, and you will have a lousy time if there is a spill.

Once a print is finished, it needs to be cleaned and cured. Keep a container of Isopropyl Alcohol (IPA) around to rinse off the excess resin and then put it under UV light to let the model dry and finish curing.

Some printers have matching cleaning and curing machines–e.g., the Prusa SL1 and Prusa CW1. These machines rinse the completed print and double as curing stations with built-in UV lamps.

Printing & Prop Considerations for Resin

There are different types of resins available for SLA printers. The standard resin is most commonly available and excellent for models, but it is also brittle. This resin is your go-to resin for prototypes and general models. If you need the part to hold up to some wear and tear or accept a screw tap, you may need to search for "tough" or "engineering" resin to make parts that will stand up to more abuse.

As a general rule, resin will work best for parts that require high levels of detail and any little fiddly models that might not print well on an FDM printer. These might be pieces like jewelry or armor ornaments used to finish a costume. If you need something that will snap together, resist impact, or stand up to some abuse, thermoplastics will serve you better.

Many resins perform better if you pre-heat the bottle to about 27-30°C before pouring it. Some dedicated resin cleaning and curing machines have a pre-heat function that will warm-up the bottle.

Cost of Resin

Resin prices vary wildly depending on the type of resin and the manufacturer. You can purchase bottles of standard resin starting at around $30 for 500g. Manufacturers generally sell resin in 500g and 1,000g bottles, but you can find much larger bottles for bulk purchases.

In general, resin is more expensive than thermoplastic filament used with FDM printers. SLA printing is just more expensive in most categories at this time. Still, SLA printers are capable of printing complex models that are completely hollow, which saves material. If you're producing small pieces for props and costumes, one resin bottle can last you a while.

Warning!

Seriously, wear nitrile gloves whenever you are working with resin. You may be someone who will not notice any skin irritation on initial contact, but you can develop a nasty allergy after repeated contact. You also do not want resin anywhere near your eyes or mouth. Getting resin residue on your hands makes it too easy to transfer that to another surface, like your face.

Nylon

Nylon is the strongest and most robust filament option. You can use it to make tough mechanical pieces, even tools. If you need something to hold up to heavy use, like a gear or screw, nylon filament may be your answer.

Nylon comes in different varieties, with nylon 618 and 645 being popular for 3D printing.

One feature of nylon that sets it apart from every other filament is it can be dyed. Nylon wants to absorb water, which can be turned into a benefit here. You can dye (even tie-dye) spools of nylon filament or finished prints in Rit's DyeMore for synthetic materials.

Printing Nylon

Nylon requires a print temperature in the 240° to 260°C range. You also need a printed bed heated to around 90°C. Nylon may require using an interference layer, like PETG, to improve bed adhesion. You can use a glue stick or something like Magigoo to help it stick.

Printing & Prop Considerations for Nylon

Like PETG, nylon filament is hygroscopic. Spools of nylon filament must be kept dry. As mentioned above, nylon can be dyed, requiring you to submerge the filament or finished print in water. If the filament is dyed – or absorbs moisture by other means – it can be baked in an electric oven before printing.

The hygroscopic nature of nylon is only a problem when the printer heats moist nylon for printing. It is alright to dye it so long as it is dried before being fed to a printer.

Also, nylon has a tendency to warp like ABS. For a successful nylon print, consider using a brim and an enclosure to help prints stick to the build plate and maintain a consistent ambient temperature to combat warping.

Cost of Nylon

Nylon is a pricier filament. The price of a 1kg spool ranges between $40-50 in 2020. Like flex filament, 500g spools are also available, so you can get some nylon filament to try for about the cost of a 1kg spool of another filament.

For models and cosplay purposes, nylon is probably something you will rarely use. Still, it is an excellent option for specific use cases. Being able to dye nylon opens up exciting possibilities for creativity.

Filament Storage

You must store your resin and filament in a dry place, away from direct sunlight, and at room temperature. You also want these materials near your printer for easy access. Your printer should (ideally) be in a warm location, so ticking the other two boxes is not too difficult.

At a minimum, consider purchasing a plastic container to keep your 3D printing materials safe from moisture and dust. Medium-sized containers from IKEA or Rubbermaid can hold several spools of filament. Purchase some silica gel desiccant beads to put in a tray inside your container. These gel beads absorb moisture from the air and change from orange to green as they take in more water. You can recharge the used beads in a food dehydrator or microwave.

If you keep the container in a location that gets some sun, look for an opaque container to block the sunlight.

If you want to be sure your filament stays dry even when printing, you can construct a fancier custom solution. There are readily available designs that hold the filament spools inside an enclosure while still allowing the spools to move and feed filament. Prusa released several designs that range from "dry boxes" for individual spools to systems that hold several spools at once[8].

Some companies have also started producing "filament cabinets" and other storage solutions to keep your filament dry. Some of these also let you feed the filament to your printer without removing them from their safe haven, like the Repkord Repbox[9].

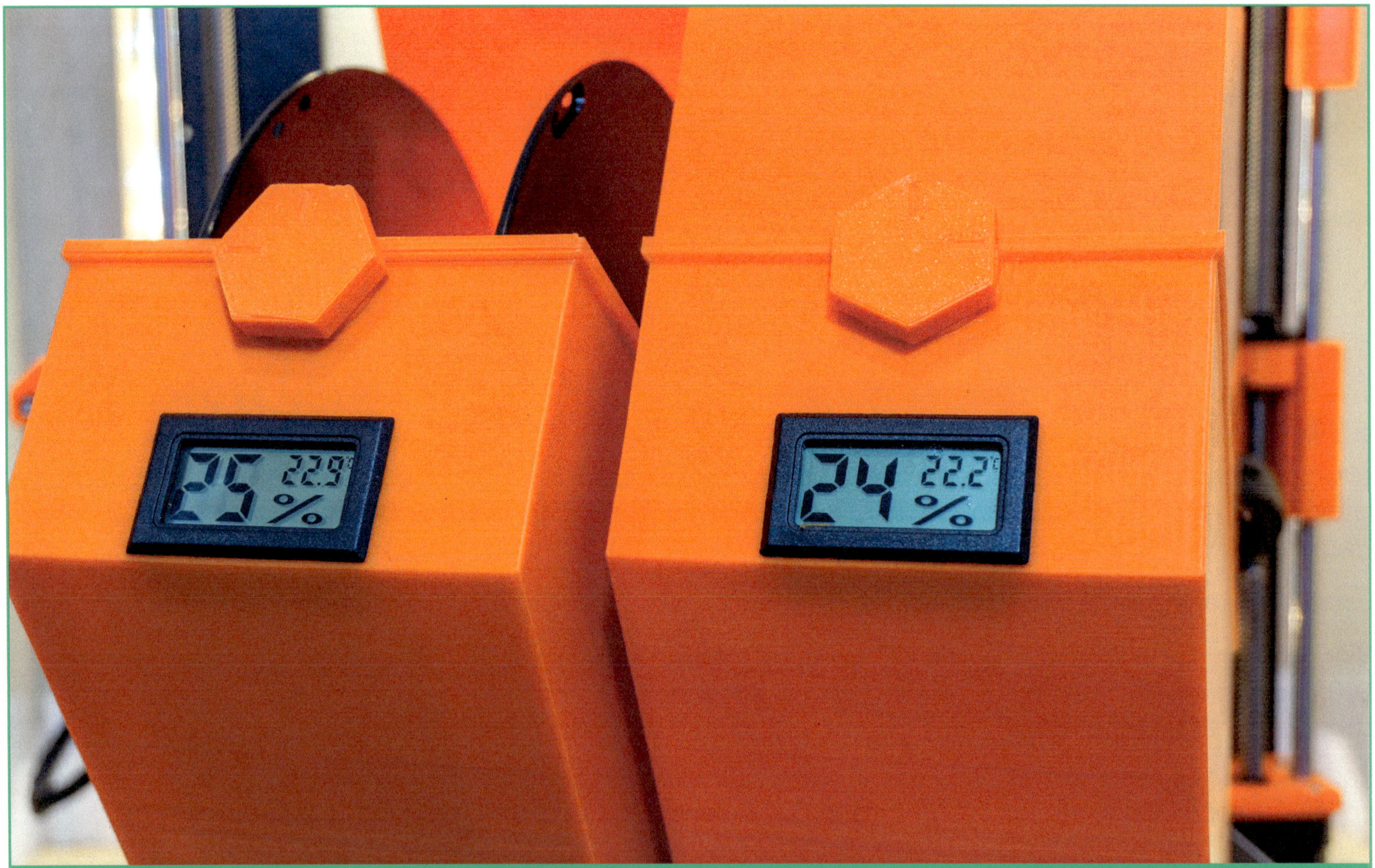

Printed Spool Boxes That Display Inner Humidity with Hygrometers

Filament vs. Moisture

Humidity changes based on geographic location, but there will be some humidity in your home or workshop. As humans, we tend to keep our average ambient room between 40-50% for comfort. While that is comfortable for us, it is better to keep filament as dry as possible–i.e., closer to 10-20%–to prolong its life.

Manufacturers generally ship filament in vacuum-sealed bags with desiccant packets. Still, spools will inevitably come in contact with moisture from the air. Sometimes this happens during manufacturing. This is really only a problem when the humidity begins to interfere with extrusion.

Filament that has absorbed too much moisture will lose its flexibility. Slightly bending the filament will cause it to discolor and begin to break. Older filament that has absorbed more water may just snap with a slight bend or twist. Any filament that has gotten that bad might break during extrusion and cause a print failure or jam the extruder.

This piece of filament came from a spool that was about two years old. A mixture of age and moisture had made it so brittle that it would snap like a dry twig if bent beyond 45°.

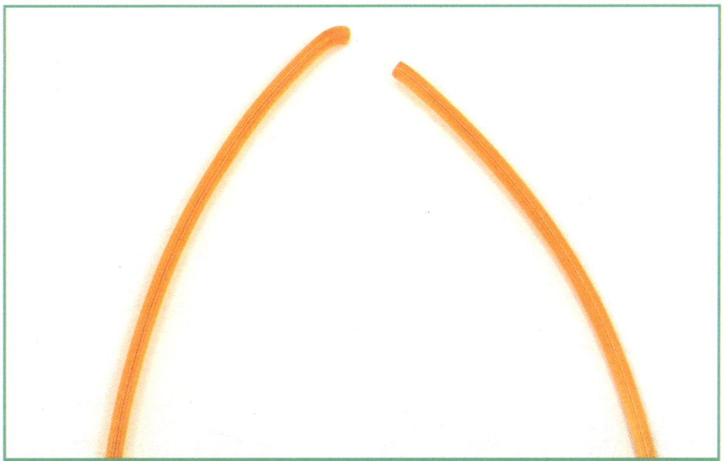

Snapped at Just 45°

Even if the filament appears to be okay, moisture can cause problems during and after printing. During a print, you might hear a popping or sizzling sound. Those sounds are a telltale sign that there is enough moisture in the filament that the water is not immediately boiled away by the hot end. This can be difficult to hear over some printers' mechanical sounds, so you might only see the moisture's impact later.

After a print is complete, you may notice uneven layers and surfaces that look unusually "fuzzy" or bumpy. These issues are not just cosmetic. Suppose you see these issues outside of the print. In that case, the interior support material (infill) is probably weak and more akin to a cobweb than a support structure. The moisture causes uneven extrusion and poor adhesion, so layers separate with a little coaxing. Those extrusion issues are even worse with infill because slicers usually default to printing infill at faster speeds. That leads to a thin, stringy infill.

It can be tempting to move ahead with printed pieces that show signs of moisture issues but be careful. The heat and pressure from sanding tend to separate uneven layers. Further, the weaker infill will not work well with screws. You may put work into a weaker print only to have it fall apart during the final assembly or while wearing your costume. The last thing you want to hear while sanding a part is a crunch.

Drying Filament

You can use an electric oven or food dehydrator to dry-out a filament spool that you suspect has absorbed too much moisture. The drying process involves heating the filament to a temperature around 20°C below the filament's glass transition point (where it begins to melt) and letting it bake for 4-8 hours. This process softens the filament just enough to release the moisture. If you use an oven, it is imperitive the oven is electric, not gas. Gas ovens create condensation instead of dry, hot air that wicks away the trapped moisture.

PLA has a glass transition point of 60-65°C, so you can heat an oven to ~45°C and then place the spool in the oven for several hours. It is tempting to increase the temperature to reduce the baking time, but that risks softening the filament enough that strands begin to bond together. It is much better to use a lower temperature and a longer baking time.

Drying filament works, but it is terribly inefficient. If you need to dry your filament before each use, you will be adding 4-8 hours to every print. It is far better to store your spools properly and occasionally dry hygroscopic filaments–e.g., PETG.

When doing this, make sure your oven or dehydrator maintains a consistent temperature. Let the oven heat up to the desired temperature and then wait. Ovens tend to overshoot the temperature and then cool down. Likewise, many inexpensive food dehydrators use coils to manage temperature and do so with less accuracy. Once you are confident, whatever you use will maintain the set temperature, insert the filament spools, and set a timer.

A few manufacturers have started offering dedicated filament dryers, like PrintDry. These devices work just like food dehydrators, except they have presets for types of filament and allow the spools to spin and feed a printer from within the device.

The following table provides some suggested temperatures for the most common types of filaments.

Material	Temperature
PLA	45°C
TPU/TPE	50°C
ABS	60°C
ASA	60°C
PETG	65°C
Nylon	70°C

Recycling Waste Materials & Being Eco-Friendlier

3D printing is not an eco-friendly hobby. Printers use a lot of energy to create plastic models with materials shipped from all over the world that is not readily recyclable. However, there are a few things you can do to be eco-friendlier.

Locate Local Recycling Facilities

Locating your local recycling center is your first step. Most filaments and resins are classified as "Type 7" plastics, or "Other." Municipal recycling centers do not always accept these plastics. Find a local facility that accepts Type 7 plastics and bring them your failed prints and scrap material. If you toss prints in with your other recycling, they may end up in the landfill.

Also, locate your local hazardous waste center. Leftover photopolymer resin must be disposed of as hazardous waste. The same applies to isopropyl alcohol (IPA), especially any IPA you have used to rinse resin prints. Resin and IPA should never be dumped down a drain or thrown away in containers.

Reduce Material Use

Be smart about how you use your materials. You can save time, money, and material–and produce less waste material–by using less support structures. Pay careful consideration to how you slice your models to reduce the need for support material. Reducing supports will improve the quality of your prints, reduce print time, and save you money.

Also, try to print fewer models at a time. It feels efficient to fill a build plate with models to get a lot done at once, but this might lead to more waste material. When you experience a power outage or catastrophic print failure, the whole plate is often ruined. In that scenario, instead of just one failed model, you have many and much more wasted material.

Failed Prints, Brims, Skirts, and Support Material Pile-up Quickly

Shop Smart

It is tempting to order a truckload of filament in dozens of different colors. Try to resist this urge and order only what you need for your upcoming projects.

It's fun to experiment with filament brands and colors but avoid ordering more than you need for the near future. Always remember you need to store all of these partially used spools of filament somewhere. Filament does not age well if you can't properly store it, so you may end up throwing away filament in the long run.

Bottles of resin actually have a shelf life. Manufacturers typically state resin has a shelf life of one year from the manufacturing date. In practice, sealed bottles can last two or three years. Resin in opened bottles and left to sit on a shelf for several months may separate and not re-mix, even with vigorous shaking.

When you know you will need more material for some upcoming projects, order in bulk as much as possible. Rather than ordering just one spool when you need it, order a few to stock up and cut down on the number of deliveries you receive.

Finally, consider the supplier's location. Ordering filament from a supplier on the other side of the world will cost more and require the shipment to travel a much grater distance.

Consider Going Smaller

Sometimes you see a model that calls to you. It looks incredible, so you download, slice, and print it. If this is a large model, you have used a fair amount of filament to turn it into a physical item that now requires a place in your home.

Before going down that path, decide if you want to dedicate the space necessary to that finished model. If you don't care about it being big, consider scaling it down to a more manageable size for display and to use less material.

Determine if Printing is the Best Option

Designing and printing various objects for your home is a lot of fun. Printing a shelf bracket means you can use a crazy design and print it in a colorful filament for some truly unique home decor. On the other hand, if you're making a reasonably plain bracket, a different material may be better. If you have the means to do so, you can make the bracket out of wood, significantly reduce the energy you use, and avoid plastic.

Consider if your project should be 3D printed or not. You can always take a 3D design, print just the first couple of millimeters, and use that as a template for another material.

Keep Prints Special

Finally, try to keep 3D printing special. An idle printer is not a bad thing! Just because your printer isn't being used right now doesn't mean you should print something. If you don't want to use or display a print, it winds up as waste material in the end.

If you want to print something just for fun or as a prototype, consider using PLA. PLA is made from annually renewable resources (sugarcane and cornstarch). You can take it to a recycling center later or place it outside. PLA is biodegradable and will completely vanish within about six months if left out in microbe-rich soil.

MADDALENA

Ghost Trap

Ghostbusters

Model by Sean Charlesworth[10]

Printed in ABS and assembled with real metal hardware

Finished with rattle cans, oil paints, and vinyl cuts

Slicer Settings

Slicers are chockfull of printer settings you can tweak. Many of these settings change as you swap nozzles and filaments. Some will vary depending on the type of object you print. This section focuses on specific universal settings and slicing decisions that have the most significant impact on props and display pieces regardless of the prints or filament you are using.

Print Orientation

One of the first decisions you make when slicing a model is how to orient the model on the build plate. Often, the best orientation is any sizeable flat surface against the build plate, but this is not always correct.

First, determine how you plan to use the model. If the part needs to hold-up under strain, orientation is an essential factor that affects strength. The printed layers stack vertically. The layer lines are similar to wood grain; parts snap more easily along the grain. If it makes sense for the part, try printing it in an orientation where the layer lines add strength–usually perpendicular to the applied force's direction.

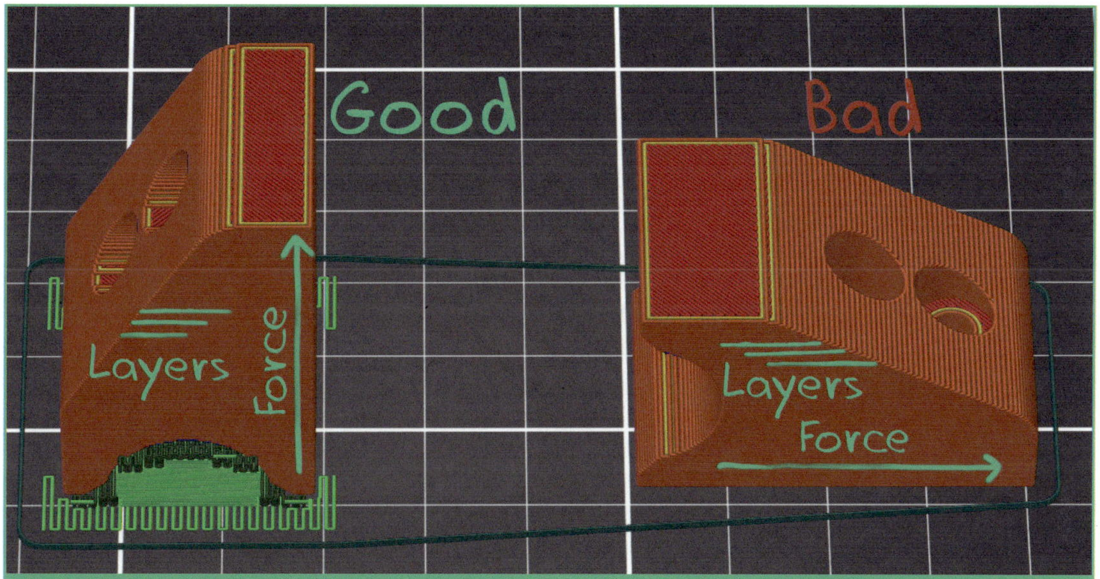

Print Orientation Strength Comparison

Take this bracket designed to support a wall-mounted Ghostbusters proton pack. The large flat surface that faces the wall is an ideal surface for the build plate. The trouble with this orientation is the pack's weight applies force down and away from the wall. Force pulls *with the layers*, increasing the chance of a bracket failure.

It is better to print the bracket upside down, so the layers are perpendicular to the wall. This orientation creates a rough surface on top of the bracket, but that section will always be hidden from view. In this case, the entire print is hidden from view by the prop, so orientation and strength easily trump the need for perfect surfaces.

Also, consider if there is a side of the print that needs to look its best. The side against the print platform shares the texture of the platform. If the platform is steel or glass, the print will have a glossy flat surface. If the bed has a textured surface, be prepared for the bed-facing side to have the inverse of that texture.

Once you have decided if you need to orient your model in a certain way for strength or looks, try to find an orientation that reduces support material and extreme overhangs.

FDM Support Material

Some prints require support material. Support material is the scaffolding that makes it possible to print extreme overhangs (greater than 45°). Occasionally, you might have a model that requires certain parts to begin printing with nothing below them, no matter how you orient the model. These models require some amount of support material.

Play with the orientation to reduce the need for support material whenever possible. Not only does this save material, but it also reduces the surfaces affected by the supports. Surfaces facing supports usually come out rougher than the rest of the model. It is preferable to print anything like an arch or a dome facing up. These structures typically require support material, which creates rough, uneven curves that are incredibly difficult to smooth.

Further, you eventually need to remove the support material. If you place supports inside a tight space, you will have a more challenging time removing them. Removal often involves using needle nose pliers and hobby knives.

Sometimes you need to play with the model's orientation. Take this slide from a model of Deckard's blaster from *Blade Runner* designed by Bill Doran of *Punished Props*[11]. The model has few flat surfaces and no obvious "correct" orientation.

The model's default orientation, pictured on the left, checks all of the boxes for strength and surface finish with a 7h 58m print time (for PrusaSlicer's default settings), but there is little contact between the model and the print bed. The small orange dot in the below image is the only part of the model that touches the print bed for the first layer. The entire model will rest on top of the support material. This is a good orientation if you are confident in your slicer and printer to produce a solid support foundation.

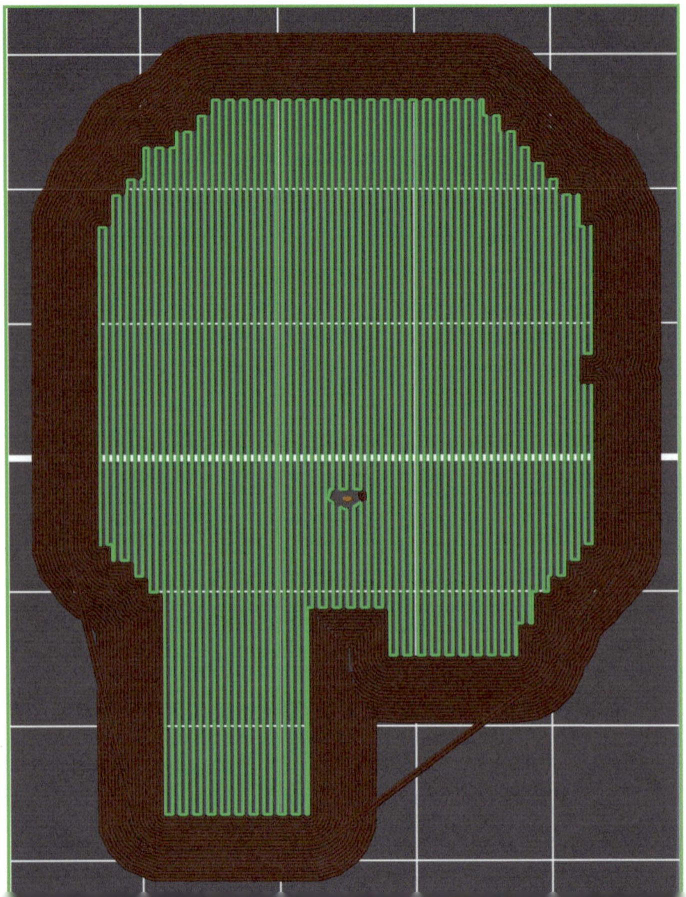

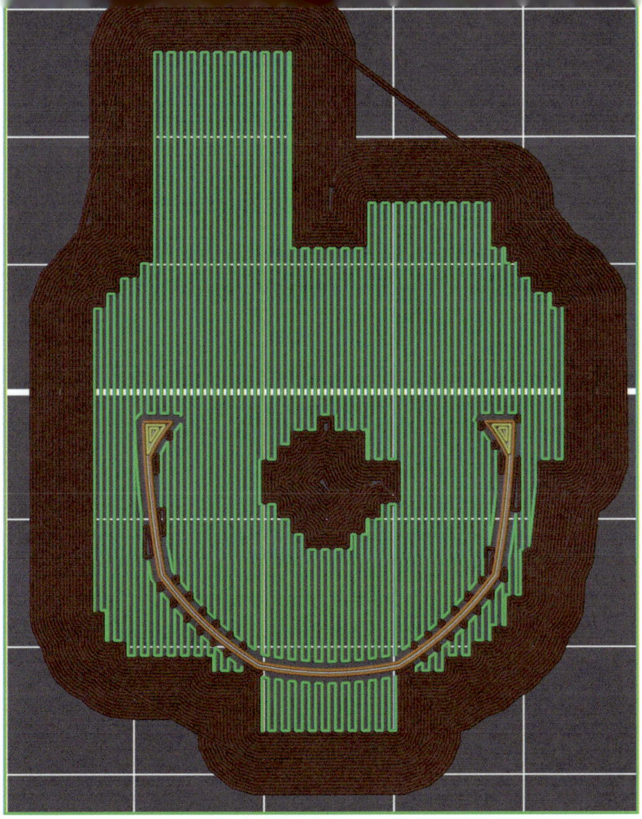

Note

Support material is meant to be light and separate easily from the model. Small printing issues can lead to the support material coming loose from the print bed, leading to a bed of spaghetti. It can take some practice to figure out what works best for your printer.

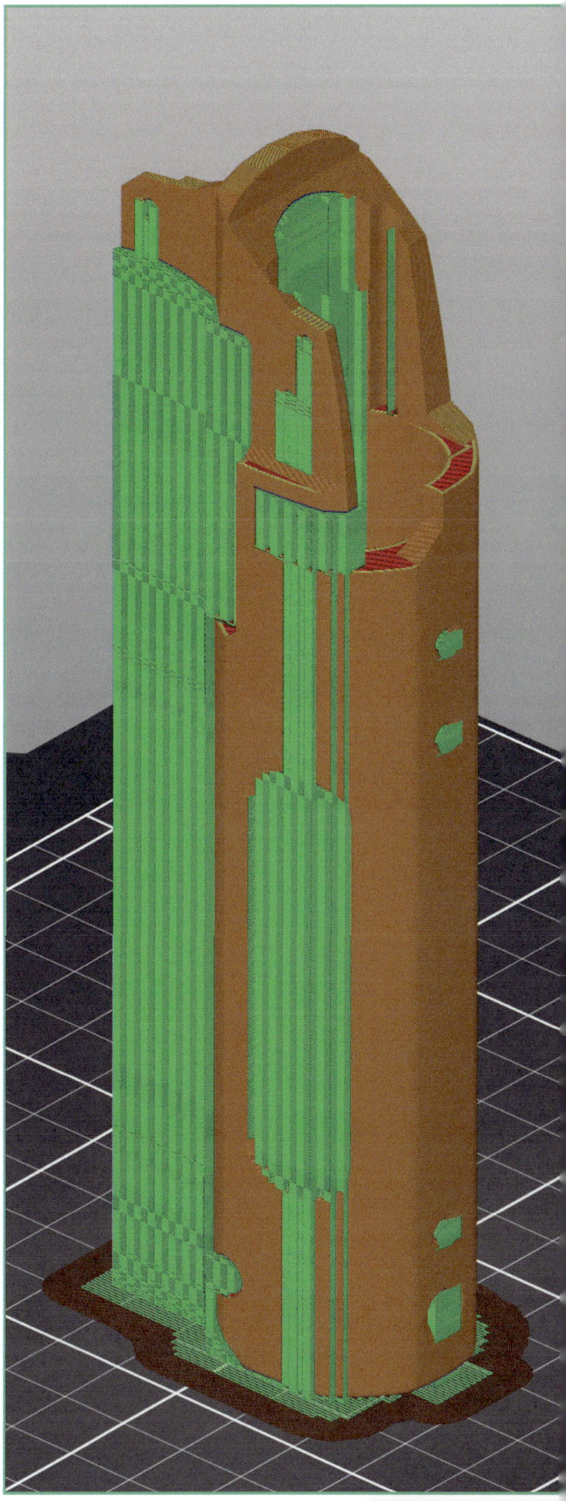

There are alternate orientations with their own sets of pros and cons. Flipping the model 180° on the Z-axis brings more of the model in contact with the print bed but adds significantly more support material and 1h 20m of print time.

The thin piece extending from the bottom of the model (the top left of the model in this orientation) caused most of this additional time and material. In this position, it's at the top of the model instead of near the print bed, so the slicer had to add significant support material–nearly the full height of the model (153.50mm)–to support it.

Both of these orientations offer some hidden benefits. If printed horizontally, that thin piece is easy to break while sanding. In these vertical orientations, that piece has longer horizontal layers

Also, the inner and outer curved surfaces will come out smoothly. They won't have any "stair-stepping" and will require only light finishing work.

We'll explore what it looks like to print curved surfaces parrallel to the print bed on the next page.

The next issue is the location of the support material. Any support-facing surface will come out rougher than the rest of your model. Support material has the most significant impact on curved surfaces.

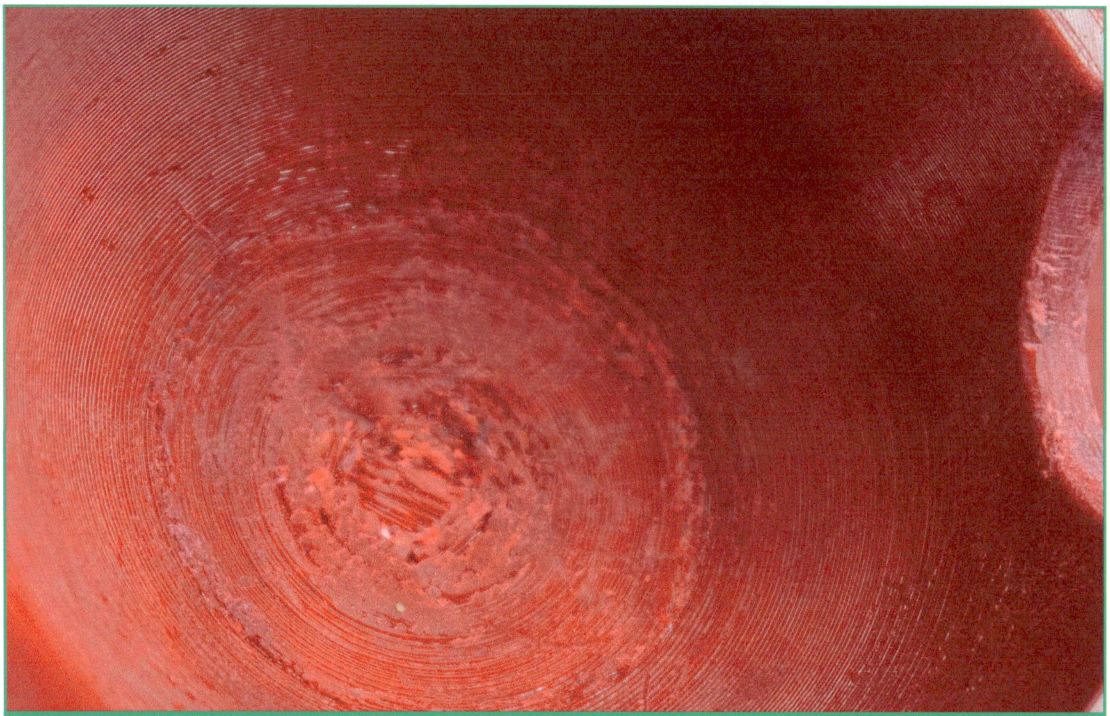

Inner Dome Printed Over Support Material

I printed the halves of this pokéball from *Pokémon* with support structures underneath the domes. The outside curves looked great, but the inside was rough and had some holes. I snapped this photo after I removed the supports and completed some light clean-up work.

Printing the outside of a dome on top of supports is even worse. The curves come out flat and ragged.

It's not only curved surfaces that suffer. I printed this part for that same pokéball replica, which required this flat surface to print on top of some support material. The texture was even but had some gaps.

Outer Dome Printed Over Support Material

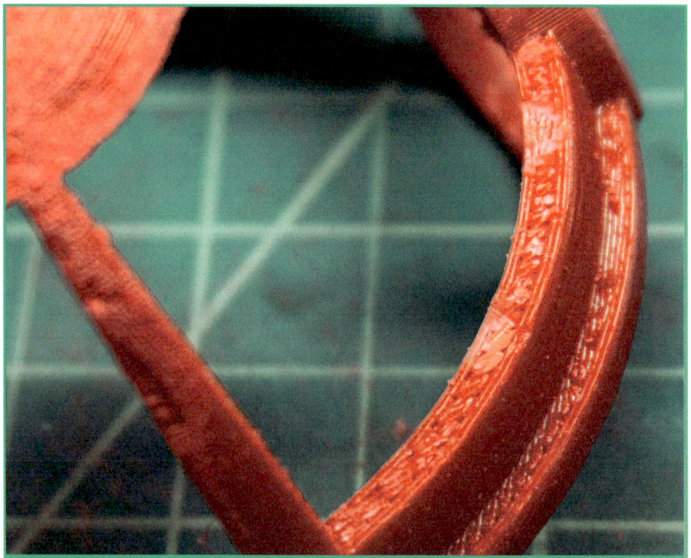

Flat Surface Printed Over Support Material

It's essential to experiment with orientations when dealing with complex structures or specific surfaces you need to be smooth. Try different orientations and support material settings. Observe how the different configurations affect print time and check layers to ensure overhangs and other tricky sections are adequately supported.

You can also place support material under only specific parts of a model. You are not limited to using automatic support placement in your slicer. It is not always correct to enable support material "from the print bed" or "everywhere." I enabled supports everywhere in the previous Blade Runner examples, and that added support material even in areas where it was not strictly needed (like the small screw holes).

You may find your first print of a tricky part does not turn out well. Learn from that print and make adjustments. As you print more complex parts, you will become a better judge of how well your printer handles overhangs and support material. You can then get better prints on your first try.

FDM Support Rafts

A raft is a type of support structure for FDM printers. The nature of the raft depends on your slicer. The raft can be a roughly printed block of filament or a layer of filament with some number of layers of support material printed on top of it. The raft acts as a pseudo print bed that sticks to the model without being too difficult to remove.

Rafts were standard in the early days of consumer 3D printing. Online resources commonly recommended rafts to solve first layer adhesion issues for small models. They did a reasonable job compensating for uneven print beds, but they wasted a lot of material.

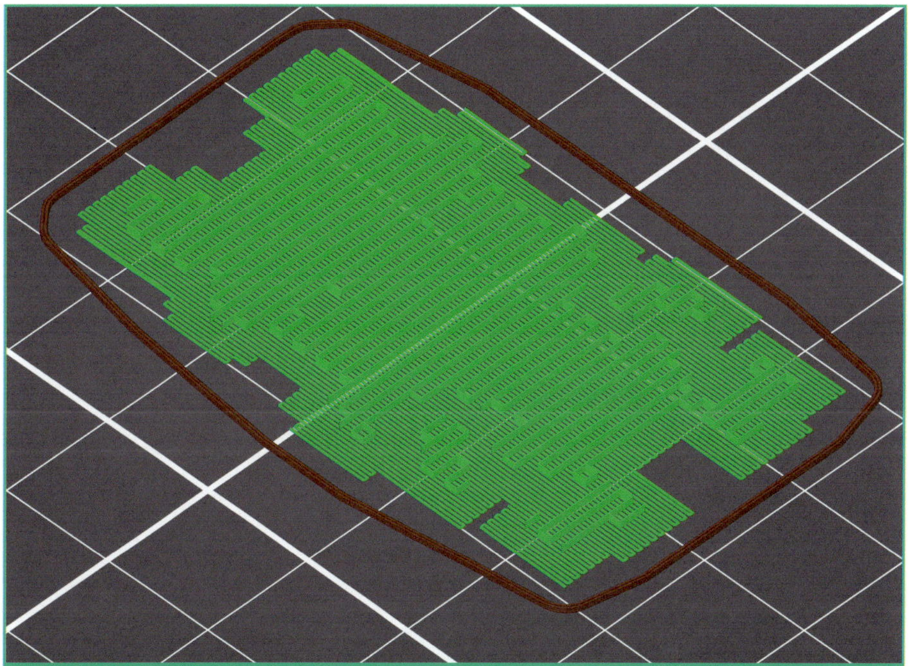
Skirt Surrounding a Raft

Today, rafts are mostly unnecessary, and you should avoid them whenever possible. Some slicers have even hidden the option to enable rafts. In PrusaSlicer, you can only see the option for adding a raft if you enter the *Advanced* mode in the print settings.

That said, you may have realized a raft sounds like the support structures in the Blade Runner examples–i.e., the model sits on top of support material without touching the bed. A raft is similar but different. Avoiding rafts does not mean avoding support structures.

Tip

If you are printing a model with curves and no large flat surfaces (like a ball) on an FDM printer, it's usually better to cut the model in half and reassemble it later. Hiding that seam will be much easier than reshaping the curves.

Brims and Skirts

Brims and skirts add extra material around your prints–extra material with a purpose.

Skirts help prime the extruder by creating an outline around the prints. The skirts do not touch the prints; they are just one or more lines of filament printed a few millimeters from the print's outer edge. Observe the skirt to catch printing issues before the printer begins. If the skirt does not stick to the bed, the print may experience issues. Loose skirts can also get dragged into the main print.

Some slicers add additional loops or layers to the skirt. By making the skirt a few layers tall, it can act as a buffer between your first few print layers and cool drafts of air that might cause the layers to cool too quickly. The effectiveness of this is questionable, but skirts prime the extruder and can work well as an early warning system for poor layer adhesion, insufficient filament flow, and uneven print beds. Many slicers default to enabling a one-layer skirt for those reasons.

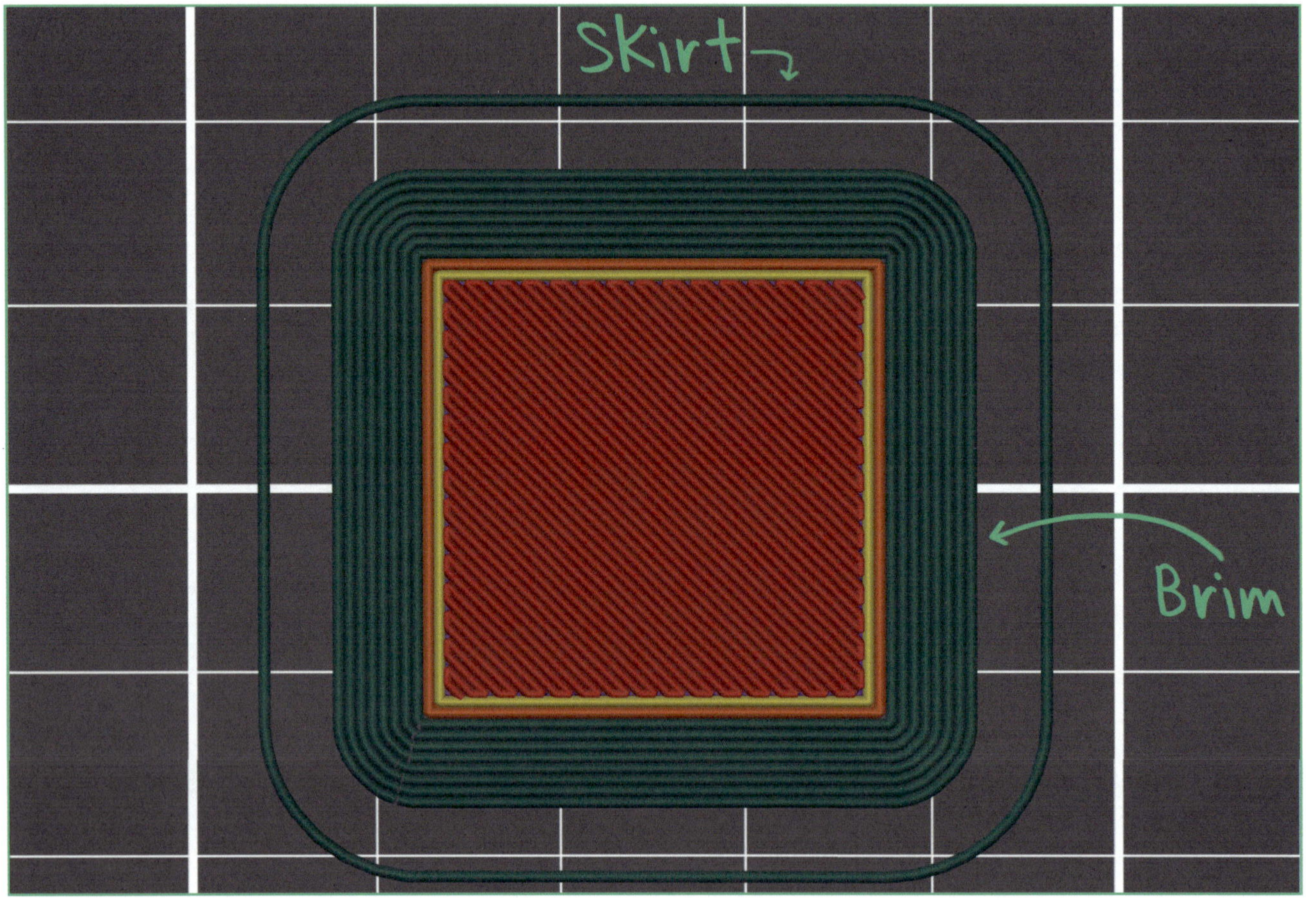

Brims are like skirts, but they connect to the outer perimeter of the print. Brims add to the size of a print's first layer to increase adhesion. They help the most when printing small objects that have little contact with the print bed. The brim helps prevent these pieces from becoming detached from the print bed or falling over.

Additionally, brims are essential when printing large models in ABS. Brims hold down the edges of the model and prevent them from lifting away from the bed. As ABS cools, it shrinks and wants to pull away from the print bed. Brims help combat the warping that can form when edges and corners curl away from the bed.

SLA Support Material

SLA printers use different support structures. When slicing a model for an SLA printer, you will typically place a "pad" below the printed object. The pad is usually a few layers thick and larger than the model. It is similar to an FDM support raft. The pad provides a foundation for the little tree structures that support the model as it is printed.

These trees anchor pieces of the model that begin unattached to any other part of the model. When planning your SLA supports, do not blindly trust the slicer to generate proper structures. Always slice the model and check for problems. This example shows the first few layers of a model. The first orange layer (the model) appears on top of a pink layer (a support stem). This arrangement should setup the print for success.

Don't go overboard with clusters of supports because you will need to remove them later. Unlike FDM supports that you can peel away with some pliers, you should remove these support trees with flush cutters. Snapping the supports off will leave behind bumps or divots on the surfaces of your model.

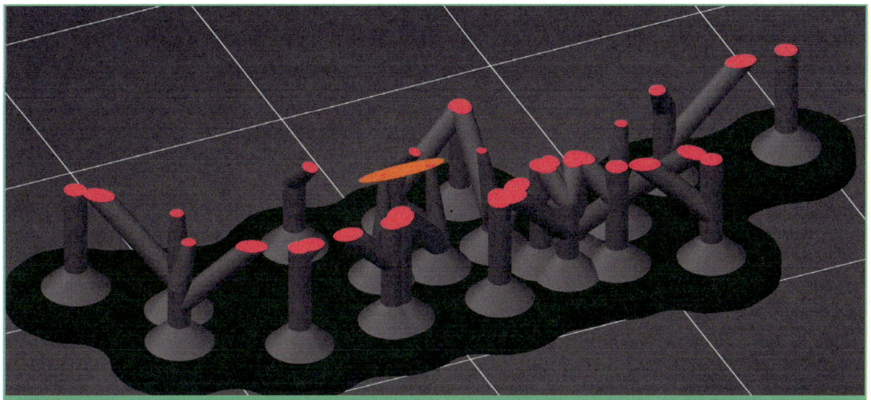

SLA Support Structure

Slicers are getting better at automatically generating support structures for SLA prints, but tend to be much more aggressive and wasteful. Take this flash muzzle for the DL-44 by PortedReality for example.

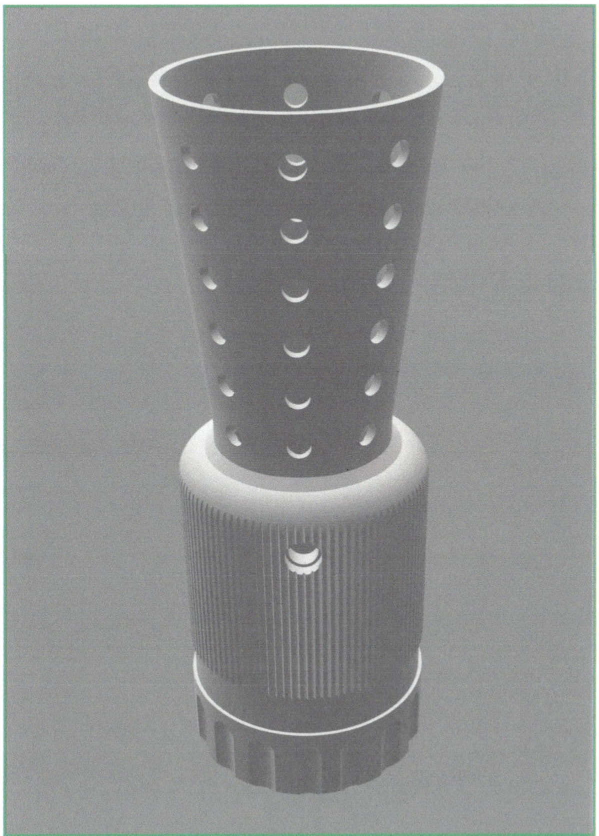

Preview of the Flash Hider

The model has no extreme overhangs or any parts that print in the air. Every layer begins touching the previous layer. That is true in all orientations. For this example, let's say we're printing the part vertically as pictured on the left.

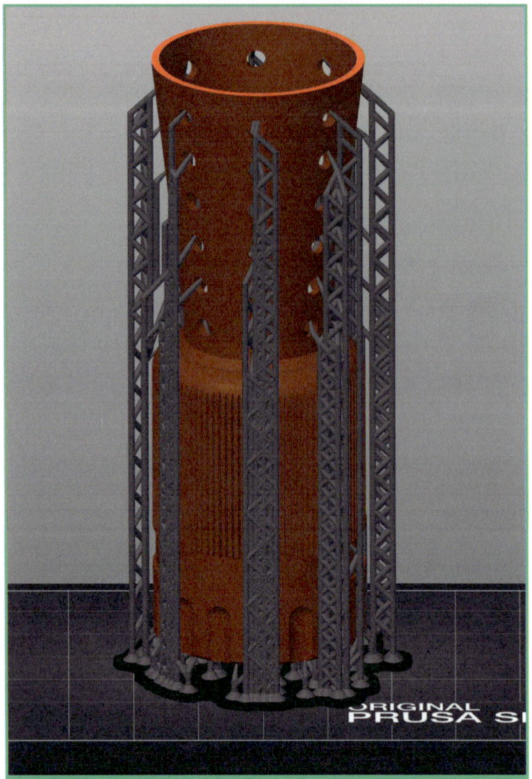

Sliced Automatic Supports

Automatic support generation adds many unnecessary support points. There are support points inside of every hole and a dense cluster of supports at the bottom. Additional supports don't increase print time with SLA, but they do waste material, and dense clusters of supports can be difficult to cleanly remove.

Manual support placement takes time, but it's worth it. When placing support points, you want to anchor the lowest point of any part of the model that prints disconnected from other parts of the model. In this case, the only face that must be anchored to the build plate is the flash hider's bottom face.

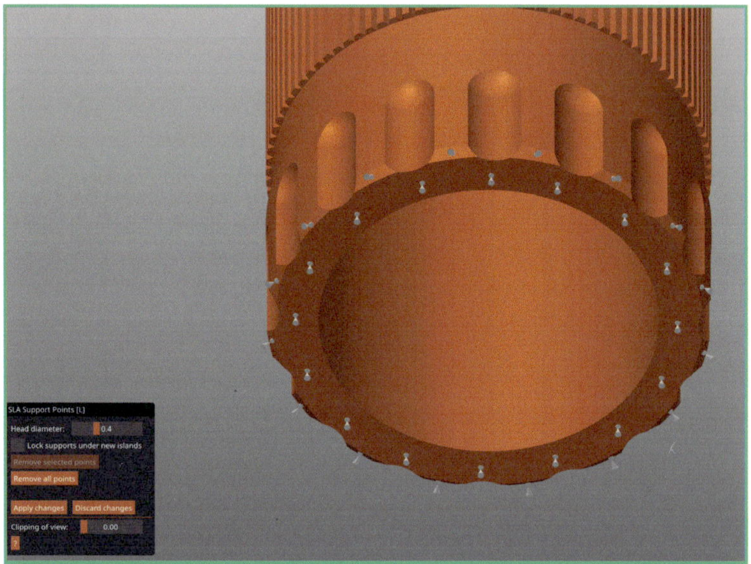

Manual Support Points

Sliced Manual Supports

Scrub through the sliced layers to check for anything that prints in the air without supports (like below). This model makes that easy, but more complex models often have some not-so-obvious spots.

Finally, check for anything that might benefit from additional support due to weight or orientation. Theoretically, this flash hider could be printed with only a handful of support points on the bottom, but I staggered 32 points in total for a strong connection.

Missing Support Point

Infill

The infill is printed inside the model to add strength and structural support. The higher you set your infill percentage in your slicer, the denser the model will be.

You do not always need infill. Printing with 0% infill will save time and material when your part doesn't need internal support material for a successful print.

Gengar Looks Empty

This Pokémon model and bust of Darth Vader both printed upright with zero infill! The models had no flat surfaces that needed infill for support. The shells (walls) made the finished print plenty strong enough for sitting on a shelf. A hollow print also gave me the option of filling the models with another material like expanding foam, sand, or resin to add strength or weight.

The darker lines are inside the model. Even printed in an opaque purple PLA, the lack of infill allowed light to pass through the model for a nice effect. With two perimeters, the model's outer layers are a mere 0.8mm thick.

You will want to use a higher infill percentage when you intend to drill into the part, use screws, or need the part to stand-up to some pressure without collapsing. Infill also helps layers stick together (more plastic in each layer improves adhesion).

If you notice layer separation with 0% infill, consider increasing the infill to 5%.

One of the common mistakes beginners make is assuming high infill creates a stronger part. That is incorrect. Setting infill to 100% would create a solid part; however, the part would still be a series of layers. The part would be difficult to crush but could always split apart if force were applied with the layers. Unless it is needed to support drilling or screws, high infill percentages are usually a waste of time and material.

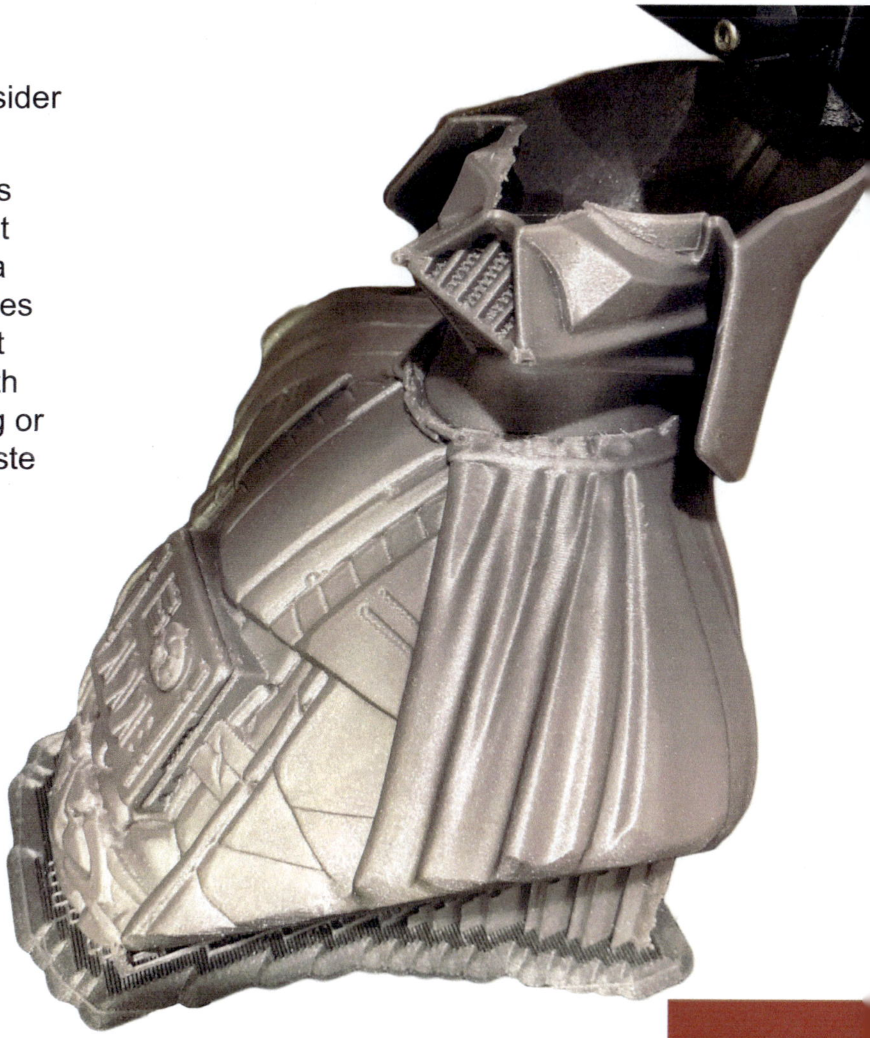

Note

For most applications, an infill percentage of 5-15% is plenty to create a strong part that looks good. I usually do not go above 20%. The trick is picking the right infill pattern.

Look to careful orientation and extra shells for adding strength. Adding perimeters to something like a screw or hook will be far more effective than just increasing infill.

Infill Patterns

Slicers offer a variety of infill patterns. There are advantages and disadvantages to each type. To showcase a few options, I sliced these boxes with multiple infill patterns at 15% density. Different slicers offer various collections of infill patterns, and some patterns go by multiple names. The following covers some of the most common options.

Simple infill patterns like lines and grid are alright for stationary models and prints you don't expect to hold weight. For more functional prints, consider high strength patterns like gyroid and cubic.

Of course, infill patterns designed to make strong prints don't always work well for prints you want to be flexible. For flexible TPE prints, slicers offer infill like concentric and cross. Gyroid is also a great option.

Finally, there are infill patterns that are mostly cosmetic. These patterns don't offer much additional strength or other benefits, but they look pretty.

In general, the gyroid pattern is an excellent all-around infill pattern for props and other models. This pattern offers superior strength in all directions, doesn't add much additional print time, and looks great (when you can see it). It also evenly supports flexible prints while still allowing them to squish and flex.

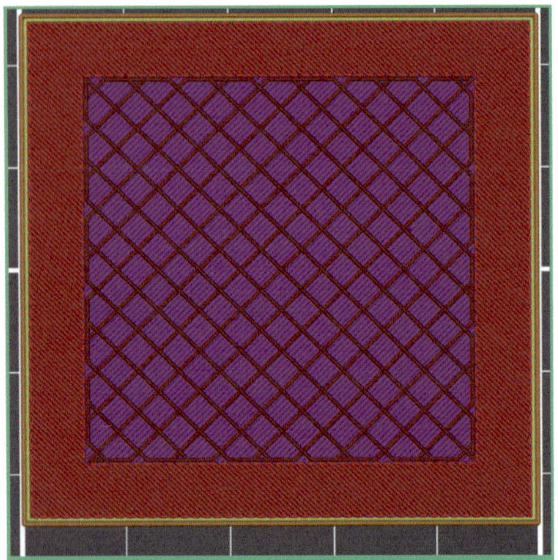
Rectilinear

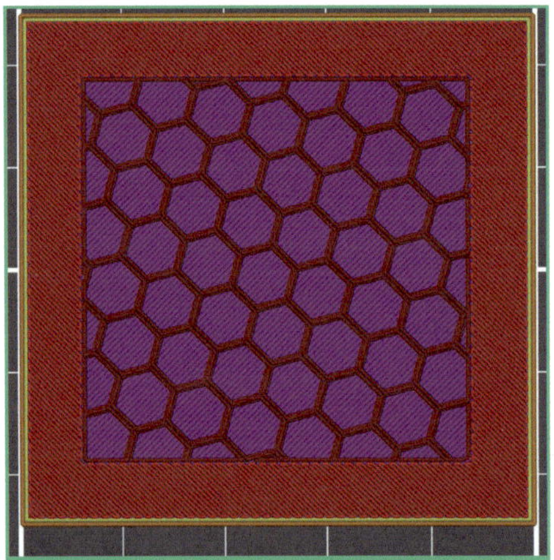
Honeycomb

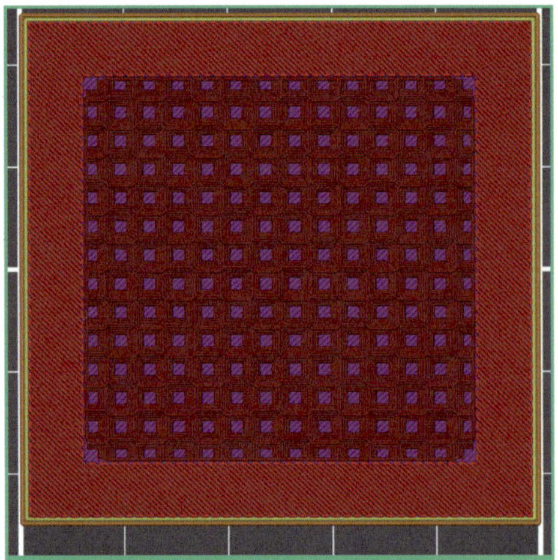
3D Honeycomb

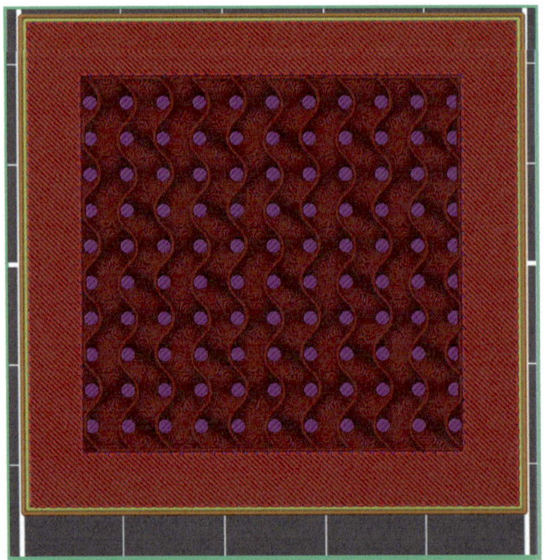
Gyroid

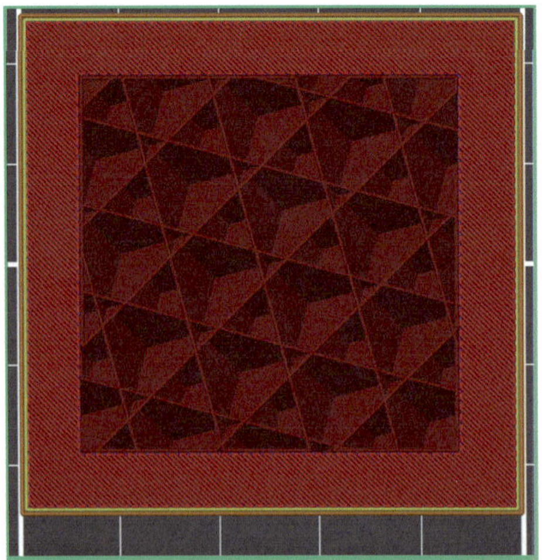
Cubic

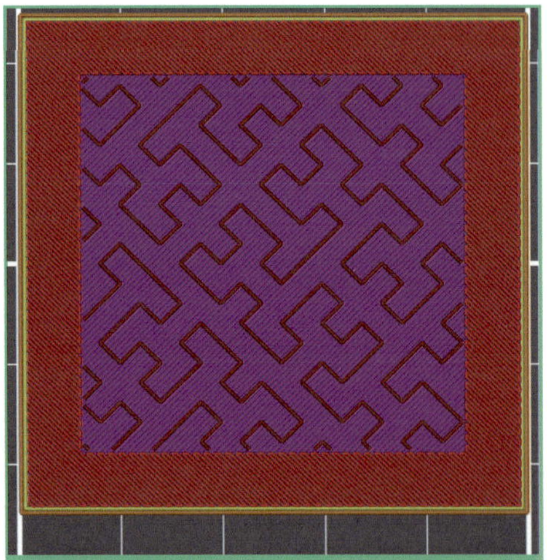
Hilbert Curve

Infill as Part of Your Prop

When you slice a model, you can tell the slicer to print zero top and bottom layers. Whatever infill pattern you select will be visible. This configuration is a cheap and effective way to transform a basic shape into something with an attractive honeycomb, grid, or mesh pattern.

You can adjust the infill density to fit your needs. You have less control over the design than if you modeled it yourself, but this trick is faster and easier. It doesn't even require modeling software.

In some slicers, like PrusaSlicer, you can insert basic shapes (e.g., box, sphere, and cylinder) of any size. (That's how I created the infill samples on the opposite page.)

If you need a quick way to spray paint a honeycomb pattern or add a neat 3D panel to some armor, insert a shape with your dimensions, set the Z-height to the thickness you want, and turn off the top and bottom layers. You can generate an instant template or accent piece without any 3D modeling.

There are so many possibilities for creative problem solving with this basic feature. Have fun with it!

Similarly, the translucent filament will also showcase your infill pattern selection. As light passes through the printed part, the infill will be visible, especially in the model's thinner areas. You can increase visibility by reducing the number of outer shells—more on that in the next section.

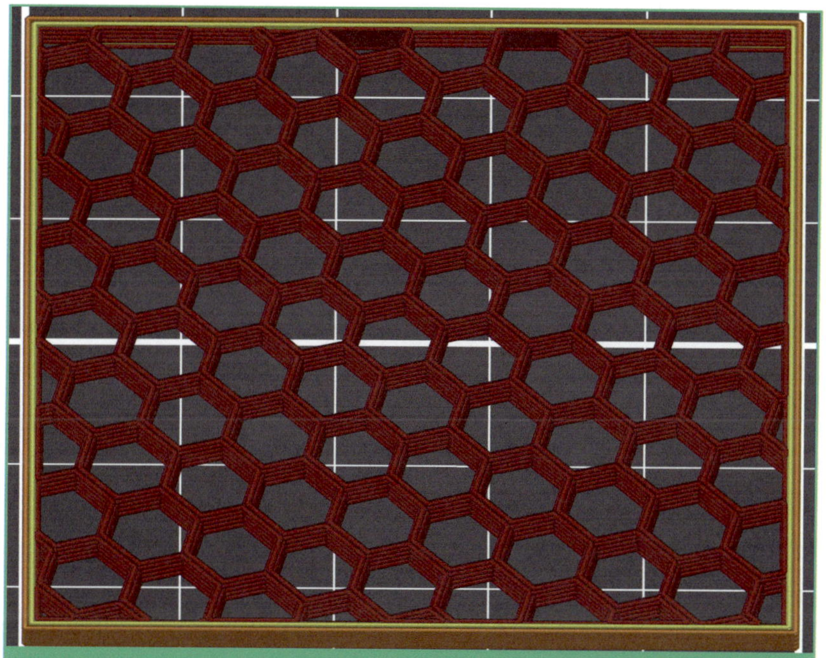

Quick Honeycomb Panel or Paint Pattern

Easy Dowel Cap or Screw Spacer

Tip

This still works even if you set infill to 0%. If you need a small pipe or spacer for a project, insert a cylinder with your dimensions, 0% infill, and no top or bottom. Without ever leaving your slicer, you now have a printable pipe. Leave the top or bottom layers turned on, and you now have a quick cap for a dowel or a button.

Shells

The term shell refers to an outer perimeter or layer. There are vertical shells (the sides of the model printed parallel to the bed) and horizontal shells (the top and bottom of the print). Some slicers drop the term "shells" and refer to horizontal shells as just the top and bottom layers.

Vertical Shells

The width of each vertical shell is equal to the width of your nozzle. For most prints, you want your outer walls to be 0.8-1.2mm thick. For a standard 0.4mm nozzle, that means you will set up your slicer to print at least two shells. If you aren't careful, you can quickly sand through a vertical shell, so consider adding additional shells if you expect you will need to remove a lot of material later.

Sloping surfaces benefit from additional shells. After the slope begins, each layer moves towards the interior of the model. The layers of a less gradual slope might print on top of the one interior shell with two shells. That would cause some loss of quality and strength. Adding a third or fourth shell uses more material, increases printing time, and gives the slope more surface area for printing. Some slicers ask you to set a minimum number of vertical shells so the slicer can dynamically increase the number of shells if it detects a slope.

If you plan to install heat inserts for screws, add an extra shell or two. The ***Mechanical Connections*** section discusses these inserts.

Horizontal Shells

The horizontal shells are your top and bottom layers. Each layer is as thick as your layer height. You control these layers' total thickness by setting how many top and bottom layers you want in your slicer.

Some slicers ask you to set a minimum number of layers and a minimum thickness. If you set up the slicer to print at least four bottom layers with a minimum thickness of 0.6mm, the slicer will automatically print two additional bottom layers if you switch your layer height to 0.1mm (0.1mm * 4 ≠ 0.6mm).

The bottom layers are the foundation of your prints. Configure your slicer to print a bottom layer that is at least four layers and a minimum of 0.5mm. The top layer sits on top of your infill (or nothing), so the first layer or two might look rough. Use at least five layers with a minimum thickness of 0.6mm.

On a well-calibrated printer, the top and bottom layers usually require minimal clean-up, which means less sanding and material removal. They don't need to be as thick as your walls.

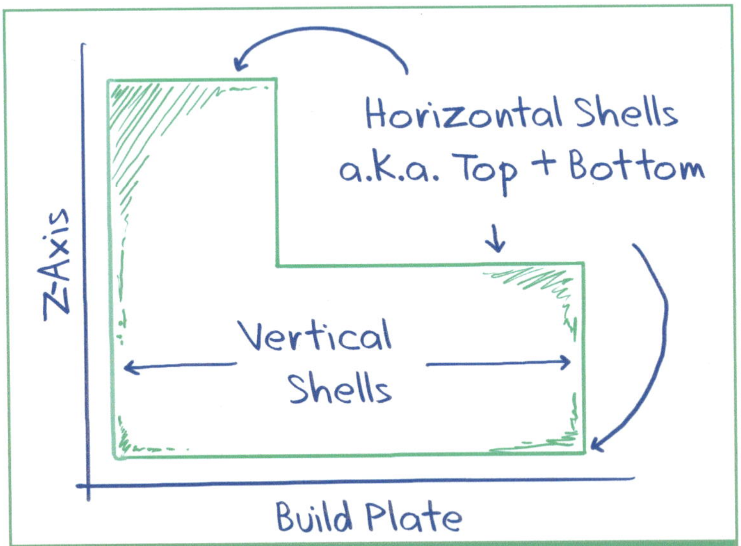

Side-View of Shells

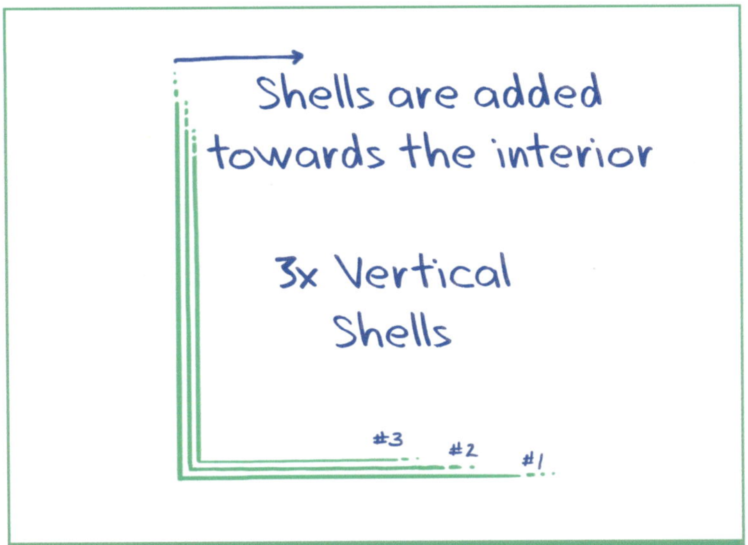

Top-View Cross-Section

Vase Mode

Vase mode is a unique setting for vertical shells that overrides many other options for shells and layers. As the name suggests, you will use vase mode when printing something like a vase.

When enabled, you may only print one model at a time. The slicer sets infill to 0%, vertical shells to one shell, and turns off all top layers. The slicer produces gcode that instructs the printer to extrude in a spiral motion. The printer will gradually raise the Z-axis as it moves around and around. With just one wall to print, the extruder never needs to stop or change direction. This produces a final print with minimal printing artifacts.

This hex box by Clockspring3D is mesmerizing when printed in vase mode. I also printed the PETG vase featured in the ***Glycol-Modified Polyethylene Terephthalate (PETG) Filament*** section in vase mode.

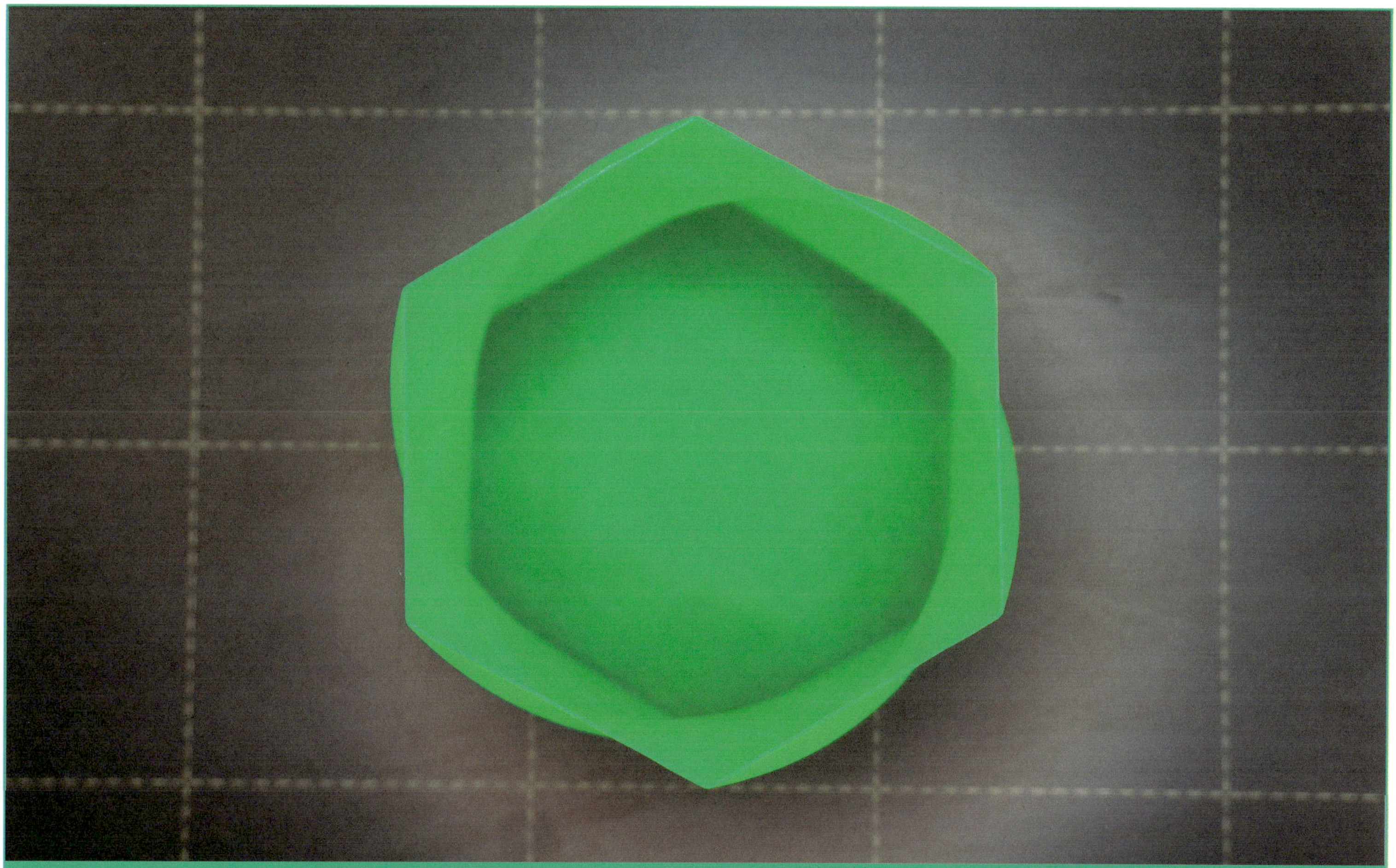

Vase Mode's Single Extrusion Walls

Note

Check your shells during a print. They should fuse together as one solid wall. Gaps indicate a problem. The printer might be under extruding because of a slicer setting or clogged nozzle.

Modifier Meshes

Some slicers support inserting meshes (models of basic shapes) that modify various aspects of your print. These are powerful tools that help you save material, cut print time, and get better prints. You can use these meshes to change shells, infill, and support settings for sections of your models.

> **Note**
>
> Slic3r and PrusaSlicer call these tools *modifier meshes*. Other slicers offer this functionality under different names and configurations.

As a case study, we'll look at Garrett Kearney's (*ChaosCoreTech* on Thingiverse and social media) model of Mjolnir, Thor's hammer. Garrett provided hollow and solid versions of this fantastic model. The head is a giant print, so it prints in halves. We will use the solid version for this example.

The hollow head is a faster print, but the finished part will sound hollow. You may also have cosplay considerations that make the hollow model unappealing. Perhaps you want to secure the handle using a bolt or dowel so your hammer will have a solid connection and still easily detach for travel or storage.

You could fill the hollow version with expanding foam or some other material. Another option is using a modifier mesh to control where you want infill and how much to reduce the print time and material cost of the solid print significantly.

I sliced the solid head in PrusaSlicer with a 0.2mm layer height, a 0.4mm nozzle, supports enabled, and 10% gyroid infill using the Prusa MK3S printer's standard profile. PrusaSlicer estimated it would take a full day to print one half of the hammer–1d 1h 39m. The slicer also estimated that the print would use about 324.12g of filament or a third of a standard 1kg spool.

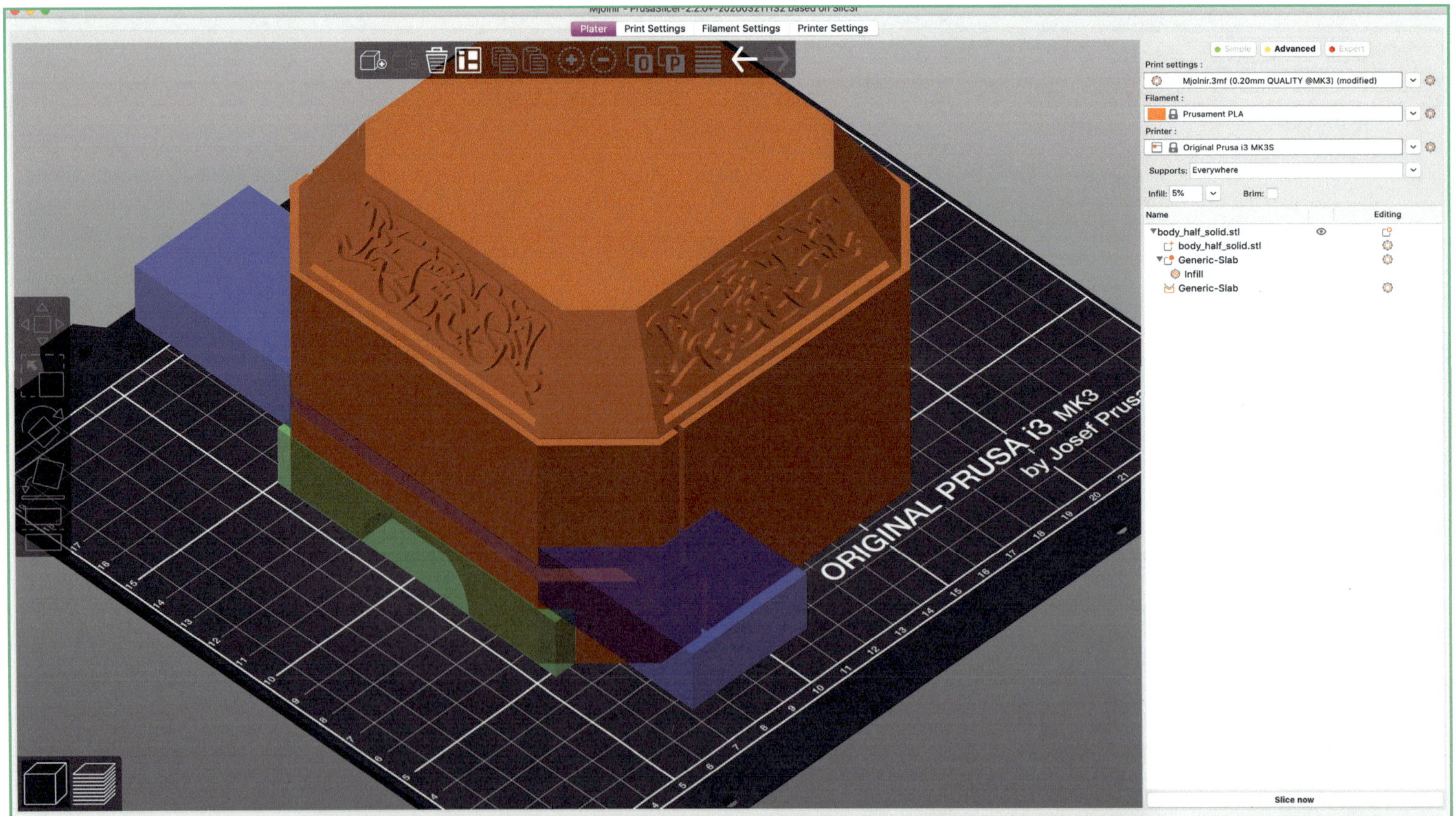

Hammer Body with Modifier Meshes

This configuration is expensive on time and materials and adds support material in places you do not want it (e.g., in the runes at the top). Assuming you were to connect the hammer with a short dowel, the hammer's head would only need dense infill around the spot where the handle attaches to the head. A solid bottom section would provide support for a dowel. The rest of the print could use less infill. Also, the model only needs support material under the small 90° overhang on one side of the model.

I decreased the global infill percentage to 5% and added a modifier mesh around the handle to give that section a 10% infill. I also added a special mesh that PrusaSlicer calls a support enforcer. This mesh added supports to the overhang without adding them anywhere else.

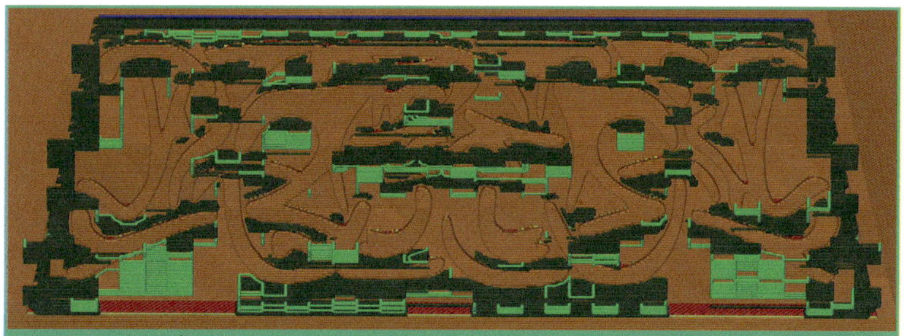

Supports Inside Runes and Details

With these settings, the slicer estimated a 16h 44m print time and a cost of 227.78g of filament–a significant decrease in time and material!

Play with modifier meshes to find the right balance for your print. What I outlined here is not the only configuration for this model. What is suitable for your print depends on your project requirements and how long you're willing to wait for a print.

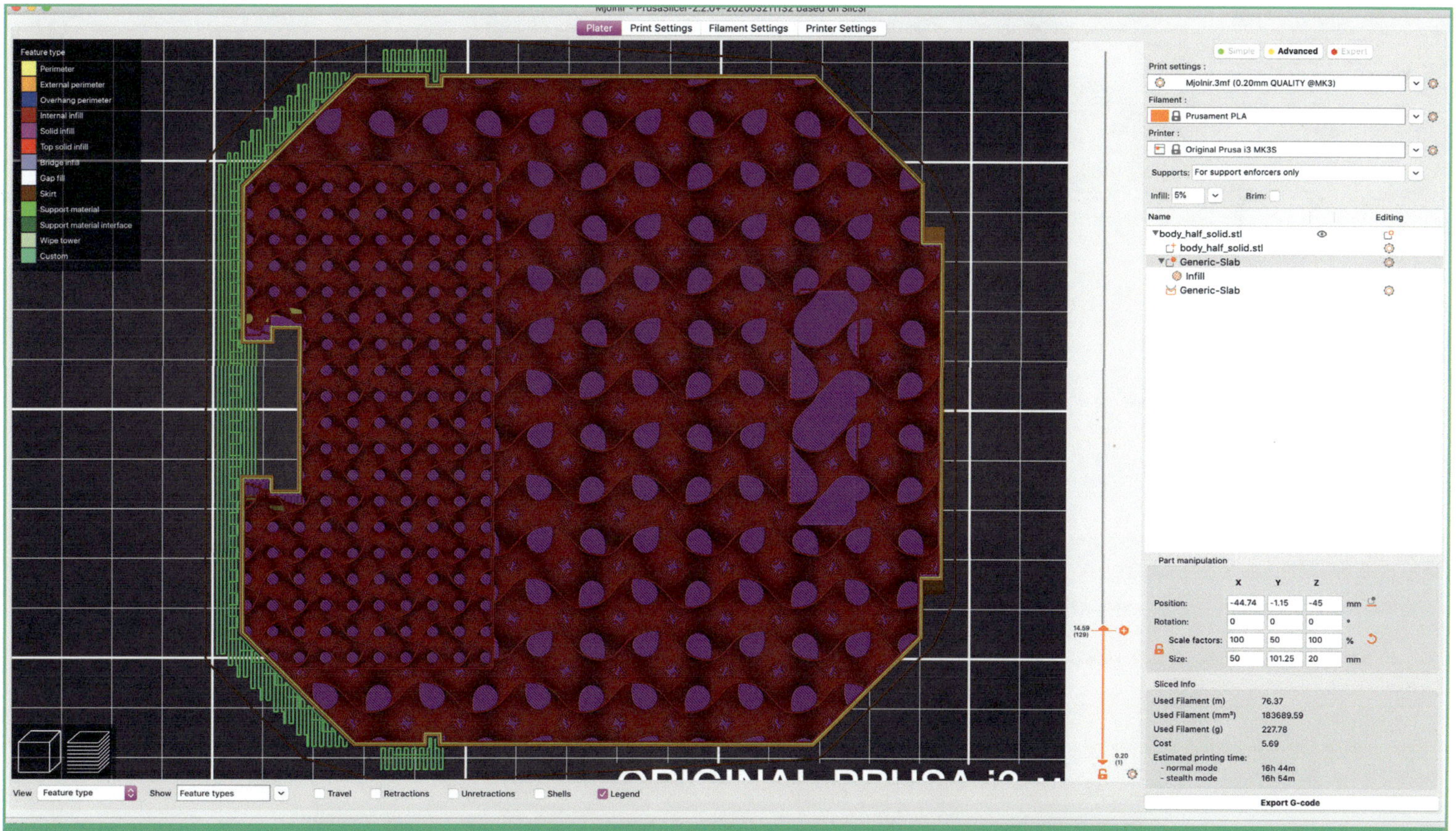

Cross-Section of Final Sliced Hammer Body with Meshes

Creative Uses for Modifier Meshes

You can also use modifier meshes for aesthetics. Like the creative uses for infill discussed earlier, you can similarly use modifier meshes. Using a mesh gives you more control over the infill and shells. With a mesh, you can also selectively remove vertical shells in addition to the horizontal shells!

Here I used a cylinder to remove the top and bottom layers and a box to create a window in a section of a vertical shell. Use meshes like these to selectively strengthen your model with extra shells or denser infill or remove layers to reveal the infill for fun effects.

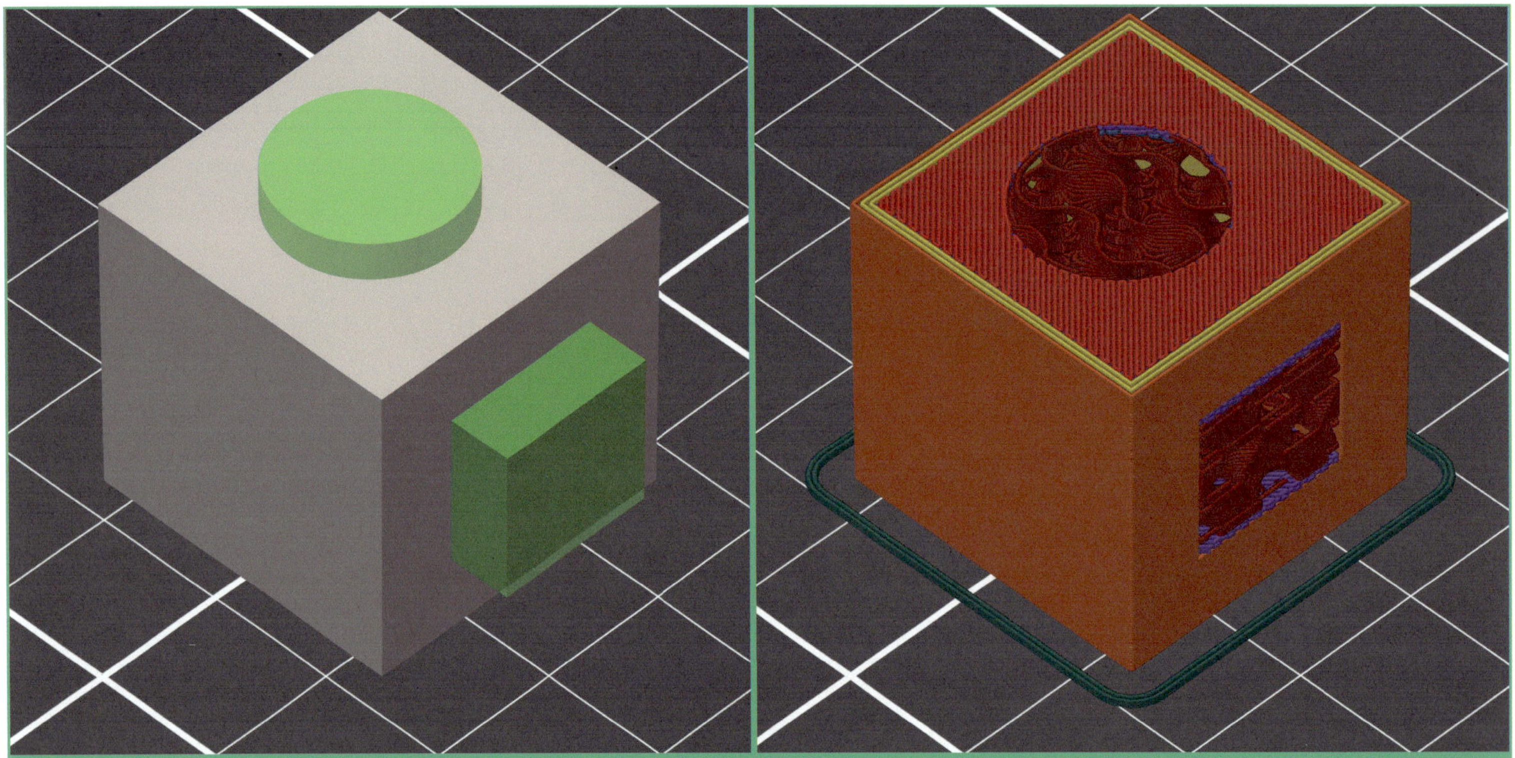

Basic Box with Modifier Meshes Removing Shells

Bust of Darth Vader by Eastman

Star Wars

Model by David Östman

Printed in PLA, sanded smooth, and painted with Alclad II Black Chrome lacquer

Custom Models

You will eventually want to print something that is not readily available online. Maybe what you can find is not precisely what you need for your specific application, prop, or costume. You could print an existing model and then alter it, but then you wait hours for a print to complete only to spend more time sanding, carving, and cutting.

The idea can seem daunting, but customizing a model–or creating your own–may be more manageable. This path is more accessible than you might expect, and it's fun!

Customizing an Existing Model with Meshmixer

When it comes to editing an STL file, your go-to tool is Autodesk's Meshmixer (*https://www.meshmixer.com*). Meshmixer is free software for Windows and macOS that can perform some useful edits to an existing model. The software might feel foreign if you are unfamiliar with 3D modeling, but it is beginner-friendly.

To begin using Meshmixer, you need to insert or import a mesh (your model). You can insert basic 3D objects as a starting point, but you will probably want to import an STL file you have and need to modify. Don't be alarmed if your imported model looks unusual. Your model may look lumpy or dented. You can fix this by opening Meshmixer's preferences and changing *Mesh Normal Mode* from *Group Normals* to *Face Normals*. I used a piece of Sean Charlsesworth's ghost trap model to demonstrate the change.

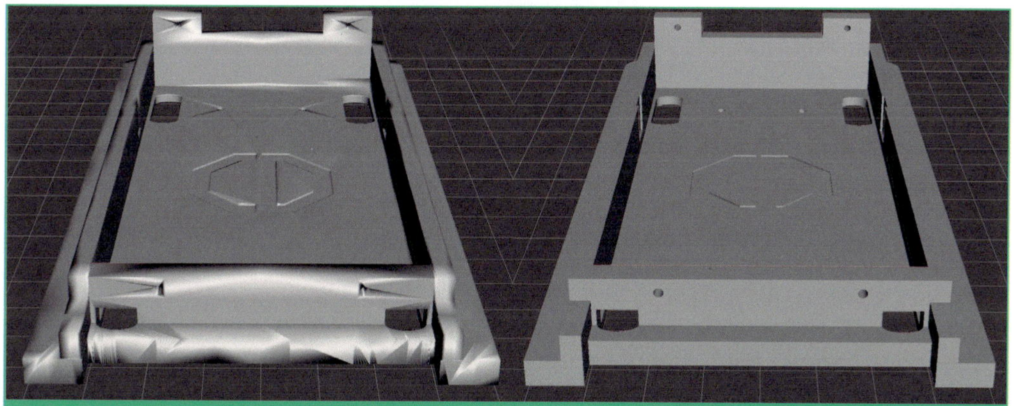

Default Display vs. Display with Face Normals

Once you have a mesh open, you can begin making changes. Meshmixer includes *Sculpt* and *Meshmix* tools that enable you to add basic forms to the mesh or sculpt changes. You can create whole models from scratch in Meshmixer.

Try-out these options to get a feel for how you might use them one day, but look to the *Edit* menu for some useful tools:

- **Plane Cut** – Cut a model along a plane at any angle. This tool creates a straight cut across the entire workspace or a selected area. You can cut models in half to fit them onto a smaller build plate.
- **Hollow** – Hollow out a model for SLA printing or to reduce material. This tool also have a button that will add drainage holes for resin.
- **Make Solid** – This tool is the opposite of *Hollow*; it makes a hollow model a solid 3D object. You probably won't find many hollow models you need to solidify, but the tool is here.
- **Separate Shells** – Split a model into two or more individual shells. You will use this after using the Plane Cut tool or when you have an STL file containing multiple models.

USING OPENSCAD

You may encounter models on sites like Thingiverse that offer *.scad files for OpenSCAD. OpenSCAD (*https://www.openscad.org*) is fascinating software that takes a programmatic approach to 3D design. The project describes itself as a 3D-compiler. Rather than using an interactive 3D modeling environment, you create models by writing scripts.

If you are comfortable writing code (or want to learn), you might try experimenting with OpenSCAD. The basics are simple. This line creates a 20 x 20 x 20mm cube centered on the origin (0, 0, 0):

```
cube([20,20,20],center=true);
```

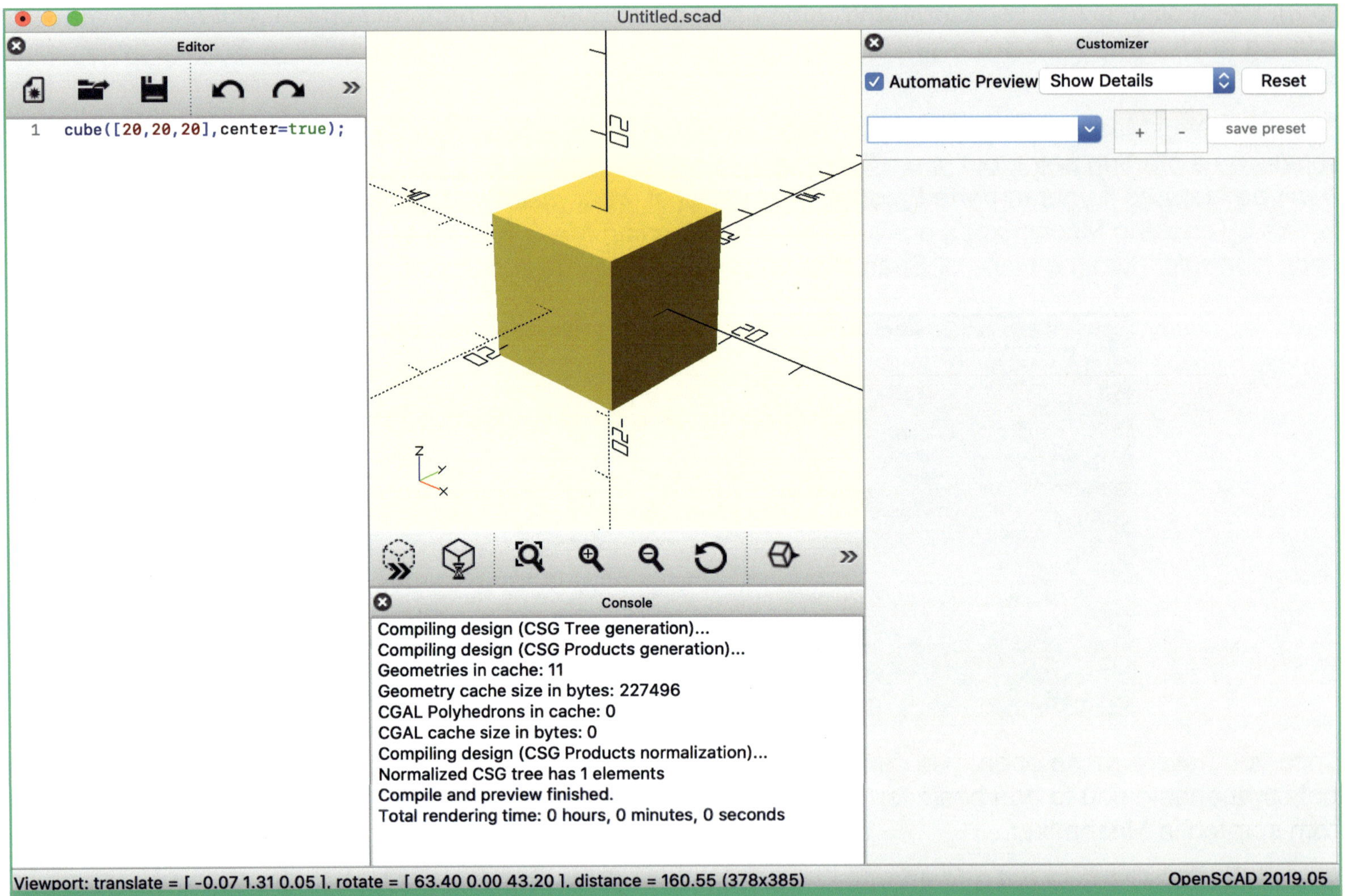

Creating a Basic Cube in OpenSCAD

You can then "translate" objects to move and reposition them. These building blocks lead to more complex shapes. With everything written in code, you can quickly modify variables in an OpenSCAD file to customize aspects of a model–variables like text, wall thickness, and spacing.

The OpenSCAD documentation has a good tutorial to help you get started. Even if you don't want to write any scripts, OpenSCAD is useful for modifying shared OpenSCAD files. For example, GitHub user MGX3D has an excellent OpenSCAD project for creating custom pegboard accessories, Pegboard Wizard (aka, pegstr)[12], which is great for organizing your shop.

Designing Your Own Models

When you think of 3D design, you might think of sculpting polygons, but many popular design programs use "sketch-based" modeling. These programs use an interface and workflow that begins with a 2D drawing called a *sketch*.

You begin by drawing a 2D sketch and then modifying your drawing with features. One of the first features you will use is an extrude. When you extrude an element of a sketch, you create a 3D body. From there, you can add more sketches and features to develop your final 3D model.

Software & Cost

Computer-Aided Design (CAD) software can be expensive, but there are many free options. Development is expensive, so someone has to pay for this software.

Free software tends to be more limited, but there are discounts for students, small businesses, and other non-commercial users.

Tip

It's a good idea to start with a sketch on paper or digitally. A Field Notes or another pocket-sized notebook is a perfect medium for capturing your ideas and taking notes. When I sit down to sketch designs, I use an iPad and Apple Pencil and a drawing app that supports layers. I can undo actions, cleanly erase lines, and hide elements (like notes and dimensions) by turning off layers.

Some applications are entirely browser-based, so you can use any operating system and basic computer hardware.

These applications come and go, change with updates, and have unique strengths and weaknesses. That makes it difficult to recommend particular software here, but a few free options have become popular with 3D printing communities.

The following sections highlight several popular options that use sketch-based design.

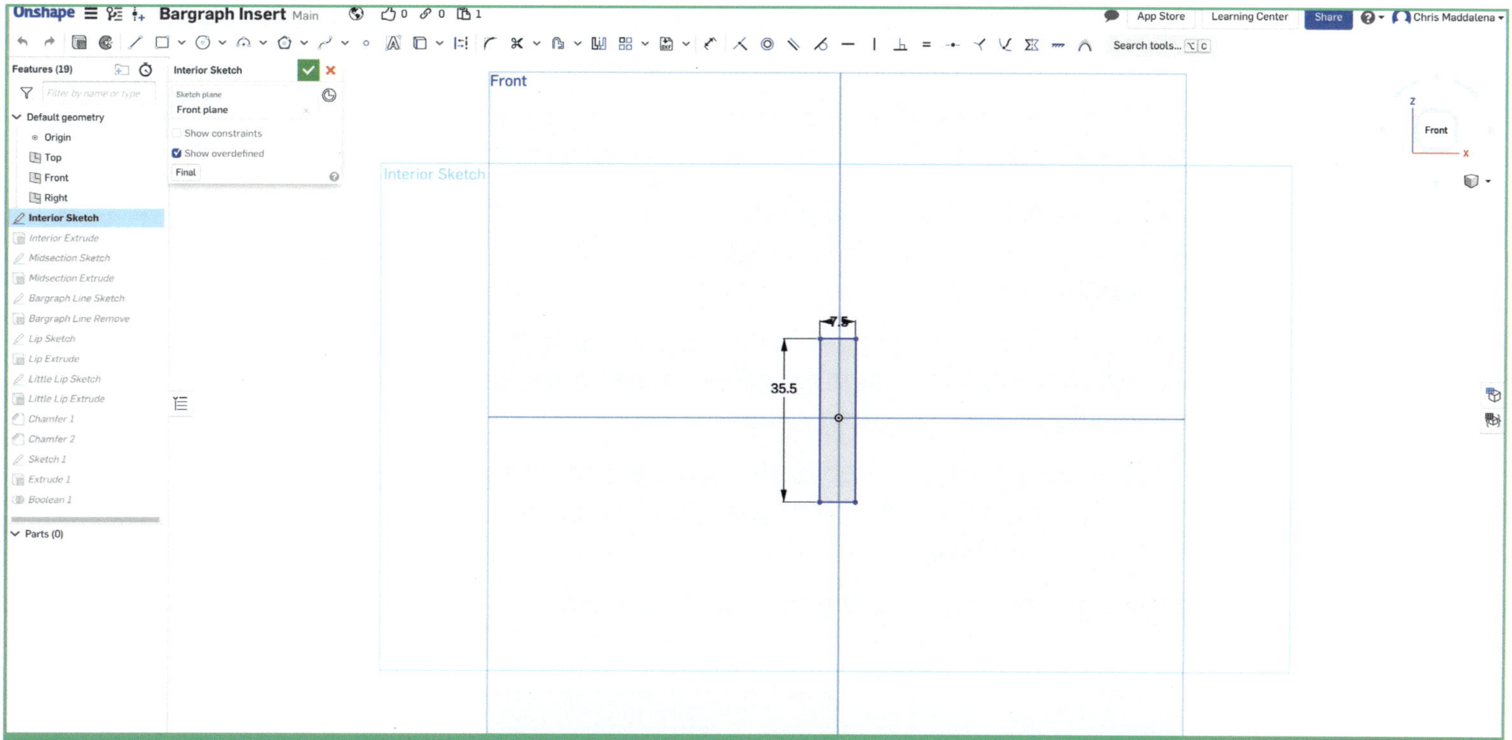

Basic Sketch in Onshape, a Browser-Based CAD Platform

Designing with Web Browsers

Many people have had success starting with Autodesk's TinkerCAD (*https://www.tinkercad.com*), a browser-based application aimed at educators. TinkerCAD comes with free projects and lesson plans to help you learn (or teach) 3D design. It also teaches basic circuit design and coding skills.

Onshape (*http://onshape.com*) and Google SketchUp (*http://sketchup.com*) are other popular, free, and browser-based options. SketchUp offers a collection of pre-made designs to help you get started with your designs.

Onshape is similar to SketchUp but focuses on designing mechanical assemblies and spare parts. With a free account, all designs are public. This is great for sharing, but it may not be acceptable if you want to keep any designs to yourself.

Designing with Dedicated Applications

Moving away from web browsers, Autodesk's Fusion 360 (*https://www.autodesk.com/products/fusion-360*) is a popular option with installers for macOS and Microsoft Windows. Fusion 360 has a free "personal use" license for non-commercial users. It is more limited than the paid version but has most of the features you are likely to need.

The biggest constraint is a limit of ten "active and editable documents." Fusion 360 stores your projects remotely, so the software needs to be connected to the internet. With a free license, you will have to deactivate designs as you hit the limit.

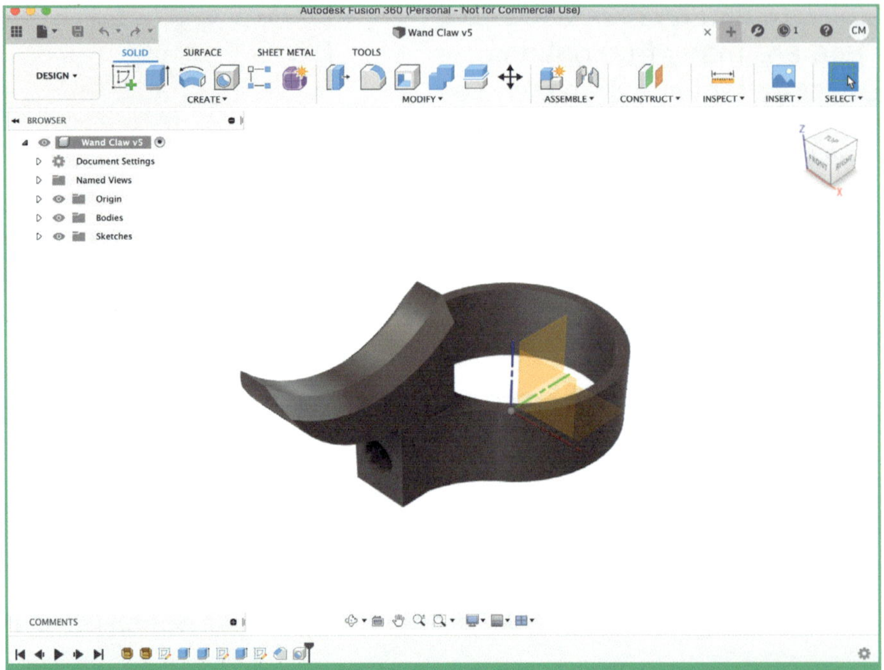

Simple Model in Fusion 360

Fusion 360 archive files (*.f3d) are easily shared so that some designers will upload them alongside their STL and 3MF files. If you have Fusion 360, you can open and customize the design.

Fusion 360 also comes with a collection of sample projects and training examples. The examples walk you through simple tasks to learn the basics. They begin with making a sketch and work up to using Fusion's more advanced capabilities.

Fusion 360 has a learning curve, but it is a robust software that strongly supports design work based on measurements and drawings (parametric design). It offers some organic sculpting, too. This book will use Fusion 360 for example projects in the upcoming sections.

Another popular option is Blender (*https://www.blender.org*), a free and open-source application made for Microsoft Windows, macOS, and Linux. Blender is a capable, professional-grade tool that excels at organic sculpts–like building with digital clay.

Which one you use depends on your goals and personal preferences.

Note

Benni of *Kamui Cosplay* has a great tutorial on how to get started with Blender on their YouTube channel, *3D Modeling with Blender for Cosplay* [15].

Design Considerations

There are a couple of things to keep in mind while working on a design concept. What works in the perfectly precise 3D world of your design space will not always neatly translate to the real world.

Always take your material into consideration. As you print more models, you will begin to understand how well your printer handles different filament materials and brands. Material shrinkage is your primary concern.

Many resin manufacturers will list an estimated shrinkage percentage on the spec sheets for their resin. FDM filament manufacturers do not do this for filament like ABS. ABS will shrink as much as 7%, but your results may vary with different brands and in different environments.

The model on the opposite page fits over an acrylic tube with an outer diameter (OD) of 22.15mm. I designed the inner diameter (ID) of the ring in the model to be 22.25mm. The model worked when I printed it in PLA. The fit was snug, but it fit around the tube.

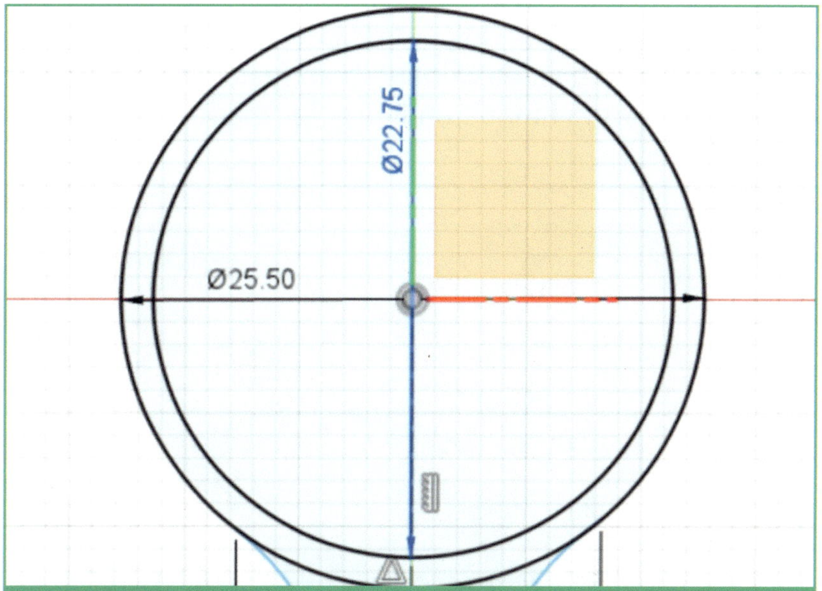

Dimensions of Ring in Fusion360

Later, I printed that part in ABS and the new ring's ID was ~21.52mm. The part had shrunk about 0.55mm or 3.39%.

I added +0.50mm to the ring's OD and ID and reprinted it. The new ring's ID was 22.20mm, reducing about 2.42%, and it fit over the tube without any difficulty. I used 0.50mm to account for a measurement error with my calipers or an "elephant foot" (a lip around the edge caused by squishing the filament on the first layer) that might have made the ring seem smaller than it was. It was a decent guess based on the results.

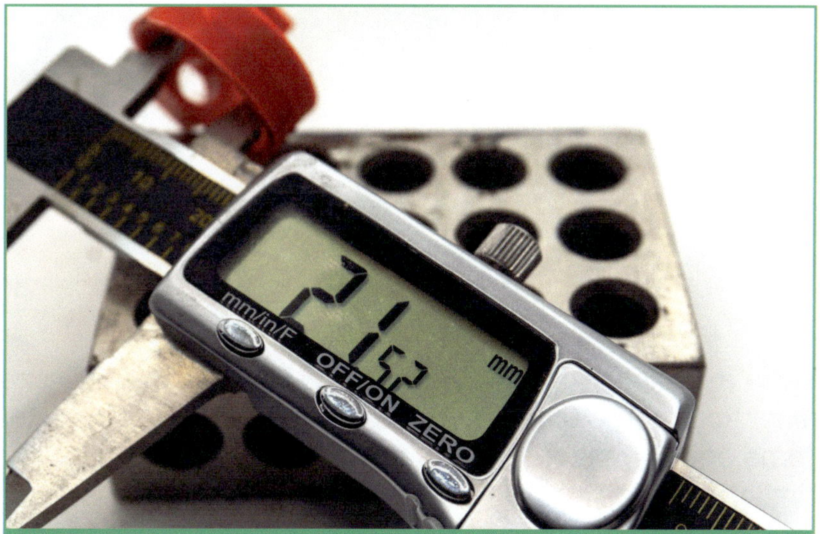

Dimensions of Ring Printed in ABS

This is another side benefit of making your own designs: you can scale your design elements to fit your material and printer.

You may see user comments on model sharing sites that say a model's parts were too big or too small. It could be the model, or it could be the user's choice of material. Designers do not always mention what materials they used during their design process. Using PLA to print a precise mechanical model designed for ABS (assuming parts will shrink) might result in pieces that fit together too tightly.

Likewise, someone might design a model with extra "slop" to accomodate imprecise printers (or vice-a-versa). This is where *tolerance* comes into play.

Tolerance is another variable related to your choice of material and printer. Your machine's tolerance determines how closely your printed parts match the intended measurements. You will get a sense of your printer's tolerances as you print models and calibrate your machine. I know my FDM printers can handle a 0.15mm tolerance. I can design a moving part with 0.15mm between it and another part of the model, and the pieces will print without fusing together.

That tolerance is tight; not all FDM printers can manage it. Knowing your printer's tolerances removes that guesswork out of the design process and helps you customize existing designs. If you want to print something that moves, like a "print in place" hinge or gear mechanism, you need to know how close you can place the walls without them fusing together. Once you know your printer's limits, you can print things like moving hinges and flexible "fabric"–like this chainmail by Agustín Arroyo (aka Flowalistik).[13]

Flexible Links Printed as One Small Sheet

Tip

You can also print onto some fabrics for some incredibe effects. David Shorey pioneered this technique. Billie Ruben and Sophy Wong have pushed it further and shared some of their successes. Check-out Billie's "floating" necklace[16] for inspiration. Billie also wrote an in-depth guide for printing on fabrics[17].

Useful Tools

When you are working with small designs, a 1mm line can look huge. It is difficult to maintain perspective as you zoom in and out and pan around a design.

Keep a straight edge with metric and imperial measurements and calipers nearby to check yourself and reset your perspective. These tools will help you check measurements as you work. I frequently pick-up items on my desk and check their dimensions with my digital calipers to compare my design's sizes to similar objects.

For checking your printer's tolerances, there are dedicated test prints available online. Angus Deveson of *Makers Muse* created this one[14]. The model has five levers with tolerances that begin at 0.5mm and step down to 0.15mm. The levers will only "unlock" and turn if they print cleanly and do not fuse with the main body.

Print a test model like this one and keep it around to help your design work.

Right Leg

Nuka Cola Caps

Fallout

Model by the author

Printed in ABS, deformed with a heat gun, and painted with Alclad II metalliclacquers and Vallejo acrylics

Example 3D Model Project

The following section walks through creating a new prop-related design using Fusion 360. You can follow along with your own copy of Fusion. This example is a simple, yet useful, custom design for a prop. There are more complex examples at the end of the book.

Proton Pack Display

Ghostbusters Proton Pack Display

Let's start with a simple design for a piece of a larger prop. In 2016, I started building a proton pack from Ghostbusters. I completed it in April 2018 with a collection of real metal hardware, acrylic, MDF, plywood, resin casts, and many 3D printed parts and fasteners.

While writing this book, I built a wall display for the pack and wanted to power it from an outlet. The trouble was I had no way to connect and disconnect a plug to the Arduino and PowerBoost 1000 boards inside of the pack that did not involve removing the pack's backplate.

To create a low-profile, hidden charging port, I decided to modify the pack's bottom plate to include a female micro USB port. This was a simple project with big utility.

Determining Measurements

Collecting measurements for a prop can be challenging but designing the model yourself offers opportunities–like adding a USB charging port. You usually try to stay true to the original's size and scale with a replica, but you can adjust dimensions as needed when you are the designer.

For this design, I re-used my 3D designs from 2016 based on Stefan Otto's blueprints published on *gbfans.com*. Stefan based these blueprints on measurements from the proton pack on display at a Planet Hollywood in Minnesota. With blueprints readily available, I skipped drawing plans and went straight to creating a new project in Fusion 360.

Note on Dimensions

When designing a prop meant to be worn, like a proton pack, you might want to scale the replica to match the wearer's size. 3D printing makes it possible to quickly and easily scale these parts for a whole family or cosplay group.

However, you cannot always scale downloaded 3D models. Scaling a model up or down affects the whole part, including things like screw holes. Creating your own models means you can scale a replica for the size of the person who will wear it without affecting the hardware you plan to use to assemble it.

Sketching the Part

My original proton pack parts were some of my very first attempts at 3D modeling. I am still not an expert, but my novice design skills have enabled me to create many different models for different projects. A finished proton pack looks complicated, but the base parts are mostly tubes and boxes of various sizes. A project like this is a great starter project for anyone new to 3D design.

The fourteen plates along the bottom of a proton pack are easy to model and print. They're just rectangles of different thicknesses. Only a few have unique characteristics that require more than one sketch to create. The bottom plate's unique feature is it's four times thicker than the others.

Here is the original bottom plate model in my Proton Pack Parts collection in Onshape.

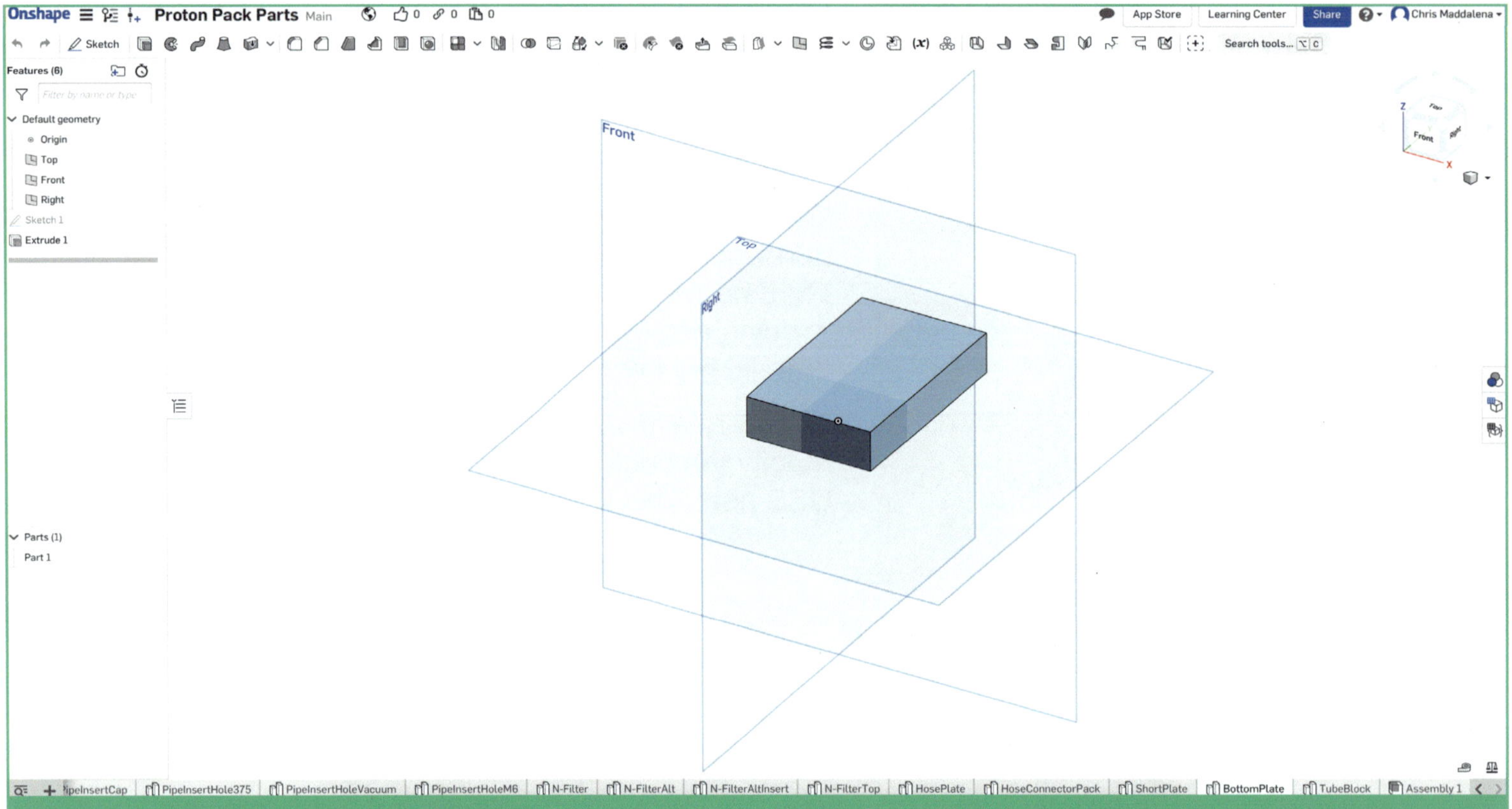

Old Bottom Plate Model in Onshape

If you look at the left side (the timeline), you'll notice only two actions:

1. Created one sketch–*Sketch 1*
2. Extruded one sketch profile–*Extrude 1*

Those are all the actions required to make a box. It's a simple design that could be easily cut from some stock material with a saw without any 3D printing. It will act as a foundation for a slightly more complicated design that would usually require additional tools, like a drill press.

Note

You can find my proton pack models on Thingiverse and Prusa Printers or view the designs through Onshape[18].

I reused the measurements from Stefan's plans and my Onshape model and made a new sketch in Fusion 360.

The base design was a simple rectangle. I created it using the *Center Rectangle* tool to lock the shape to the origin (0, 0, 0). I set the length of the sides using the *Sketch Dimension* tool (press the D key).

These two actions fully *constrained* the rectangle– i.e., locked it to the center coordinates and 39.24 x 63.50mm.

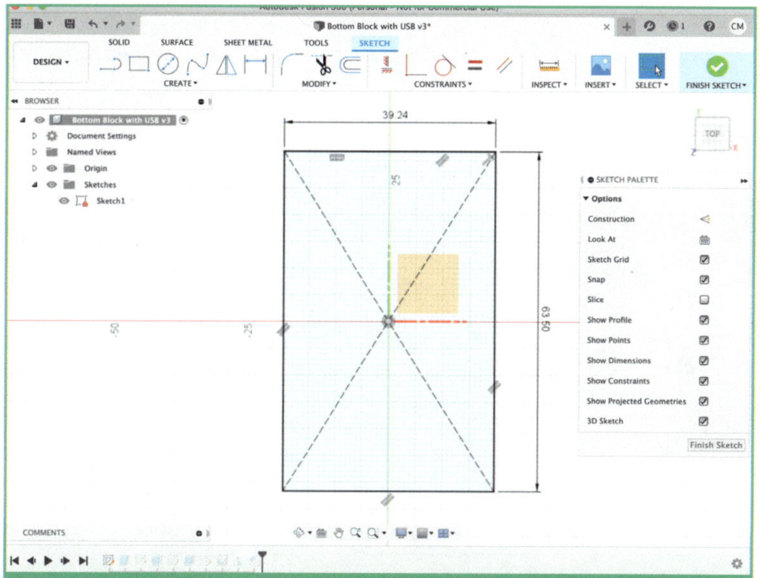

Sketching the Basic Shape

Note

In these sketch-based design programs, blue lines usually mean the lines are unconstrained and can move. The lines change to black once constrained.

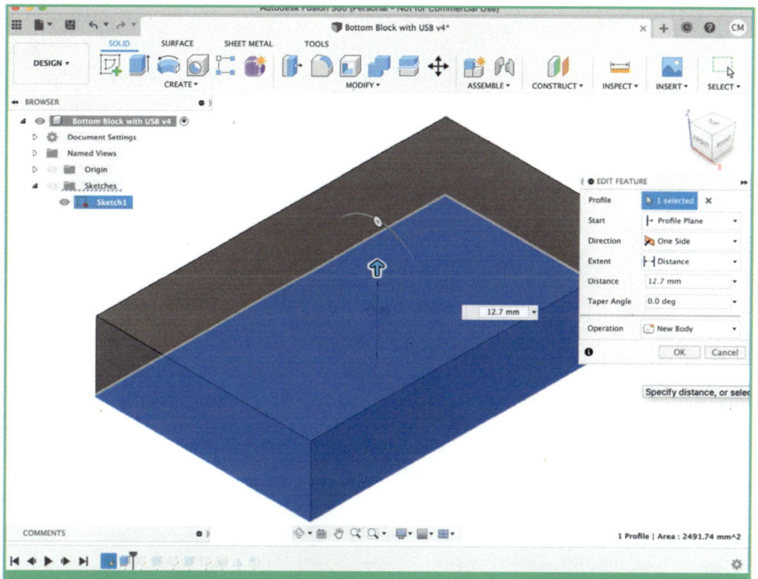

Extruding the Sketch Profile

The *Extrude* tool (press the E key) transformed the sketch profile into a 3D block. The final object was 12.7mm thick and a reasonably accurate model of the bottom plate of a proton pack.

Now the part needed some custom modifications for a female micro USB port. The cut-out could have been part of the original sketch, but it would have made extruding the part and managing the sketches confusing.

Micro USB extension cables are cheap and easy to find thanks to the popularity of Raspberry Pi and Arduino boards. Manufacturers make some designed to be attached to surfaces with screws, like this one.

I measured the port with my digital calipers to determine how wide and long the main body was and how far the screw holes were from the edges.

The USB Extension for the Plate

Tip

Billie Ruben has made some great posters[19] that discuss the differences between model-making programs and tips for designing with 3D printing in mind.

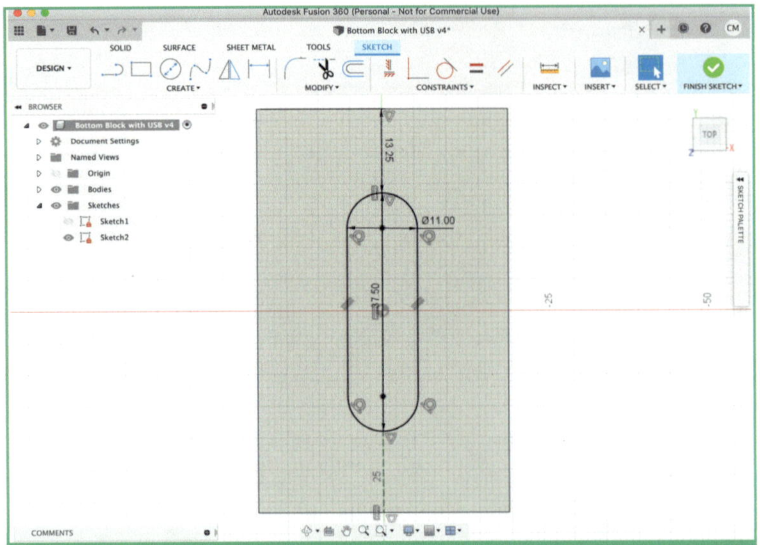

Sketching the USB Slot

With these measurements, Fusion 360's *Slot* tool placed a shape with the properly-rounded corners. I added 0.5mm (approximately +5%) to the measurements to account for ABS shrinkage.

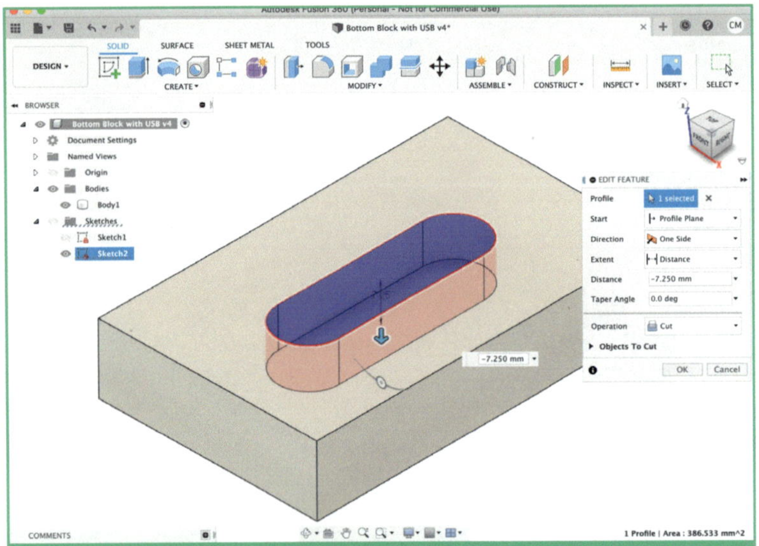

Cutting the USB Slot

I extruded this sketch into the block to cut a 7.25mm deep slot (the thickness of the USB port's mounting points). Extrusions can also remove material when set to the *Cut* operation.

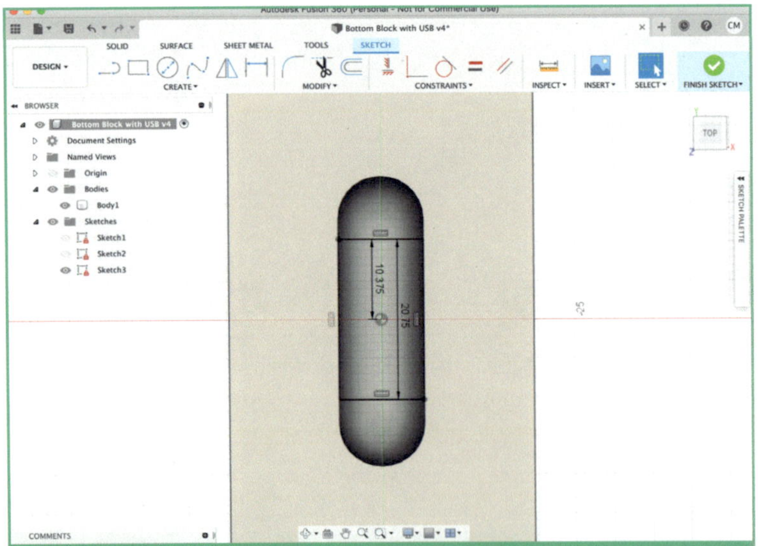

Sketching the Cut-out for the Cable

A third sketch at the bottom of the slot planned the last cut operation. This cut allowed the USB cable to exit the back of the block and back-end of the USB port to fit in the slot.

This sketch was another basic rectangle. The USB port's back-end tapered to a smaller rectangle, so the model could have mirrored that form. In this case, the block and the cable's back-end would be hidden, so a straight rectangle cut was much more manageable.

The model had one final last sketch for the screw holes. I planned to use M3 threaded inserts that called for 4mm holes. I added 0.2mm to these measurements to account for the ABS shrinking.

When padding measurements for shrinkage, use percentages instead of a static value. Adding 0.5mm to these measurements would have been too much for these smaller dimensions. The addition of +0.2mm increased the hole's size by 5%, on par with increasing 10.5mm to 11mm as done earlier.

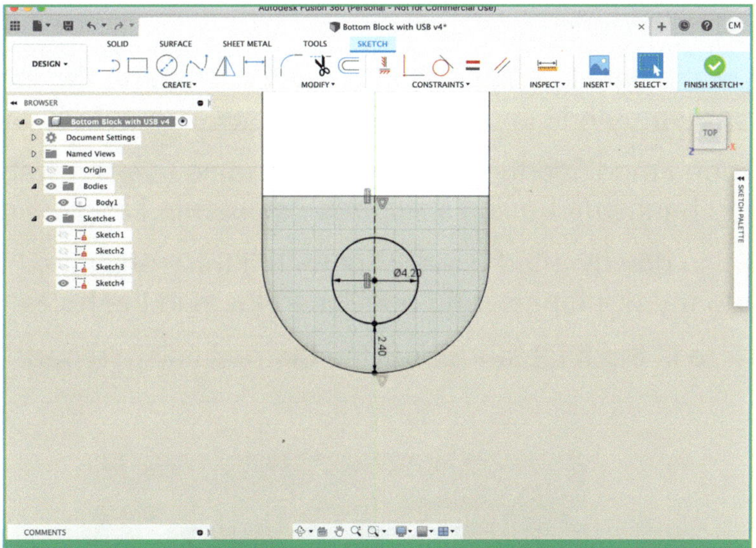

Planning the Location of the Screw Holes

The sketch profile acted as a visual guide for me while using Fusion's *Hole* tool. Another cut operation would have worked, but this tool offered more control over the cut. I chose to create a straight hole that was 5mm deep (just enough for the inserts).

The *Hole* tool can create countersunk holes for different types of screws and even tap threads. Depending on the hole selected, Fusion can automatically set parameters based on a library of screw types (e.g., M3). These options aren't always useful for 3D prints because of material shrinkage and printer accuracy when printing features like fine threads. Still, it can be helpful for planning and tinkering.

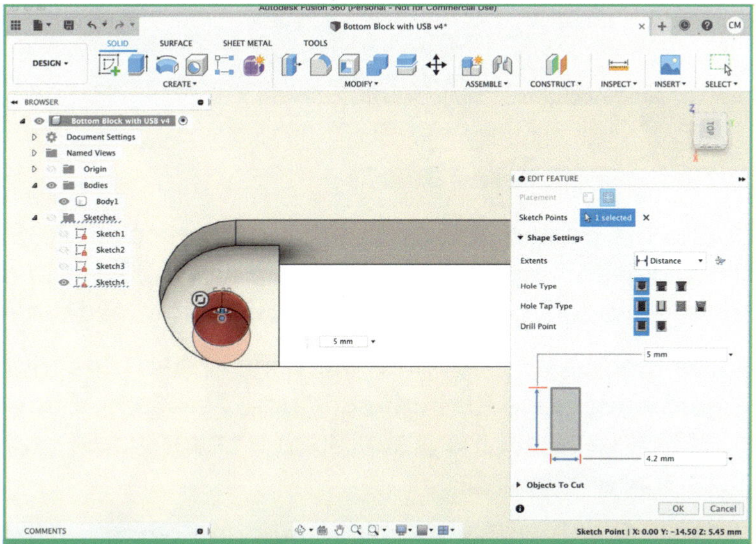

Placing a Hole for the Left Screw

The last step was to mirror the new hole to the other side of the model. Rather than manually duplicating sketch elements by hand, it's often better to use the *Mirror* tool. This tool replicated my selected sketch elements and feature (the hole). If I ever changed the original hole or sketch, the mirror operation would automatically update the copy.

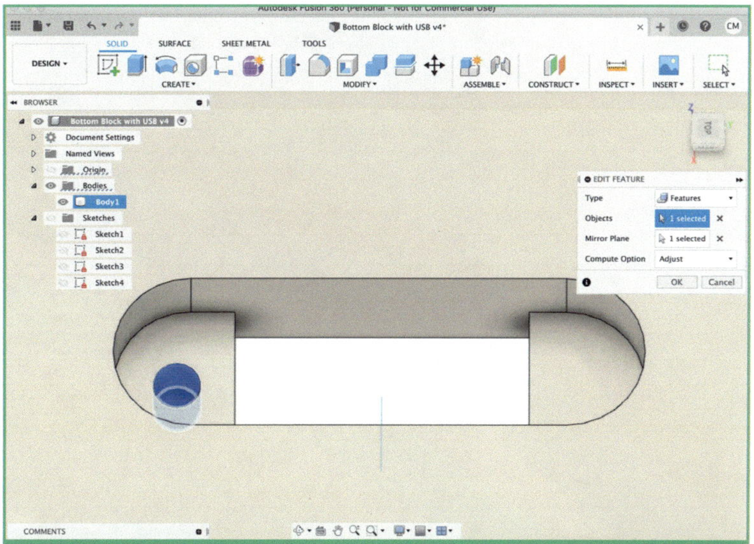

Mirroring the Screw Hole

Note

Fusion 360 can now import models from McMaster-Carr's library of screws, bolts, nuts, inserts, and more for checking size and fit.

Finishing Details

For the final touch, I added a small chamfer to the inner edges near the screw holes. This feature allowed for a smooth fit in case shrinkage caused a collision with the USB port's back-end.

Then I modified the model's appearance to see what the finished model might look like. Pressing the A key brings up Fusion's *Appearance* menu. I applied a matte black appearance to the model.

I also downloaded a model of an M3 threaded insert from Adobe's model library. I inserted two of them into my workspace and positioned them in the holes. I gave those bodies a polished brass appearance.

Here is the final render next to the real finished part.

Assembling the Final Pieces

During the design process, I printed the block in ABS and discovered my shrinkage estimates had been about 0.1-0.2mm off. I extended the slot 0.25mm to get the measurements used in this final design. This change made it possible for the USB port to slide into place with only slight resistance.

Some models require more than one or two tries to get right. That trial and error is all part of the process. I could have gone through additional iterations. For example, a deeper slot would allow for a USB cover to sit flush against the plate. Such a cover would protect the USB port when not in use and keep the plate screen accurate.

My pack is a blend of *Ghostbusters* and *Ghostbusters 2*, so I had already taken a few liberties with my design. I chose to lean into this modification and dress-up the port like it was a part of the pack's design.

The Final Bottom Plate

Buried Fossil

Animal Crossing: New Horizons

Model by DNA Designs[20]

Printed in opaque white resin, painted with Vallejo acrylic paints, and weathered with heavy body acrylics

Finishing 3D Prints

You will usually need to perform some finishing work to get the surface you want or prepare the print for painting. Even if you do not paint the finished print, you may want to clean-up the print to make it look its best. This clean-up process involves sanding, a lot of it.

Test Fit Your Parts

Before you start sanding, it is good to test fit your pieces. The last thing you want to do is realize the part is the wrong size after you invested hours into sanding and painting it. Even if you designed the model and are sure of the dimensions, the printed part may have shrunk, or screw holes might require some clean-up.

For example, this piece for a proton pack took two threaded brass rods, two M4 screws, and a brass elbow. If the holes were too small, overtightening any of the metal hardware could have caused cracks or split layers. Likewise, if the holes were too large, the hardware would have fit too loosely and fallen out.

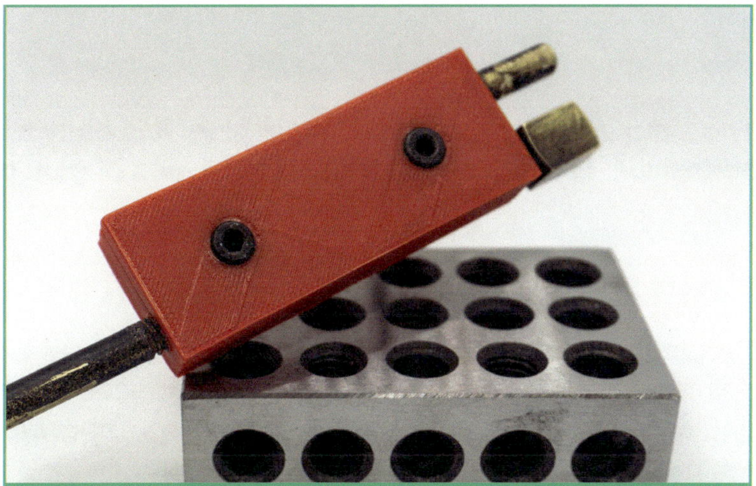

Test Fitting the Hardware

If a fit is too tight, it might cause problems later during final assembly, too. The plastic will flex and cause cracks in your paint job. Friction caused by cutting threads or a screw head rubbing the edge of a hole can also chip your paint. In this example, the M4 screws fit snugly inside of two holes. If the screw heads rubbed against the holes' sides, they would claw at the paint and cause chips and cracks that would have affected the edges where such damage would be visible.

If you use screws, a drill, or a tap to cut holes or threads into the plastic parts, clean-up the holes with a sharp blade or deburring tool. The screws should fit snugly without requiring much force. If you need to exert extra force to drive the screw, there is a heightened risk of stripping the threads or causing other damage during a later step in your build.

I cleaned and tapped a hole in this print using an M3 tap. This zoomed-in photo shows the threads and some burrs in need of clean-up.

Burrs Created by Cutting a Thread

Gluing Printed Parts

This Group Really Sticks Together

Once you are happy with how your parts fit together, it is time to consider partially assembling them. There is little reason to sand or paint parts that will later be hidden under other pieces. A partial assembly of your parts at this stage helps you sand parts evenly, fill seams, and avoid unnecessary work.

If you want to keep your parts separate to make any sanding or painting easier, just loop back to this step when you're ready.

The simplest attachment method is glue. It is easy to bond plastic parts together with a wide range of adhesives. I recommend having a few types of glue on hand for different applications.

Surface Preparation

Regardless of which glue you use, it will work best when you have a bare plastic surface. If you need to paint the parts before gluing, mask the surfaces to be glued with some frog tape (yellow painter's tape), and don't remove it until you finish painting. Otherwise, sand or scrape the painted surfaces to remove the paint or primer before applying the glue.

If you apply glue to a painted surface, the bond is only as strong as the paint's bond to the plastic.

You may also need to abrade the surfaces–i.e., give them some *tooth*. If your surface has been sanded and polished to a high shine, there is a strong likelihood that your glue won't stick to it because there's nothing for it to grip. Once cured, you may be able to peel away the glue.

Once your surfaces are free of paint and sufficiently roughed-up, wipe everything down with some IPA to remove dust, dirt, and finger oils. You want the surfaces clean, so the glue can do its work.

Picking a Glue

You want to match your glue with your materials. This is easy with 3D printed parts. Anything you are gluing probably won't be flexing much, and your parts will evenly (more or less) expand and contract with temperature changes. However, this is a book about cosplay and props, so you'll probably want to mix your 3D printed parts with other materials.

Many glues are compatible with plastic, and 3D prints are porous, so they work well with most adhesives you are likely to use for cosplay. The only glue that you should avoid is air-drying polyvinyl acetate (PVA) glue–like Elmer's school glue and Mod Podge. PVA glue works great with paper but will not do much for you with 3D prints.

Tip

PVA glue won't hold your 3D printed parts together, but it's absolutely compatible with plastic. You can apply PVA glue to surfaces before painting to give them interesting textures.

Cyanoacrylate (CA) Glue

Glue with a Kick

CA glue–you may know it as super glue–is an excellent choice for most 3D prints. I recommend the Loctite and Bob Smith Industries (BSI) brands of CA glue. This glue cures quickly and creates a strong bond between your plastic parts and many other materials (e.g., EVA foam, leather, wood).

You can choose from a few different viscosities for CA glue. They range from thicker gels to liquids that run like water. Try the different consistencies to find what works best for you. I prefer the glue that BSI calls #2 Insta-Cure and Loctite's "control gel."

BSI also produces two other products you will want on hand: Un-Cure and accelerant (or "kicker"). The Un-Cure compound melts the CA glue so you can separate glued pieces or clean-up squeeze out. It also helps get CA glue off of your fingers.

The accelerant causes the CA glue to cure almost immediately on contact. You can apply some of this liquid to one part, glue to another, and they will bond within moments on contact. Try not to get this on your hands!

The accelerant comes in a spray bottle. Keep some of it in a small squeeze bottle for precise applications. Squeezing out a drop or two is enough to get into the seams of your glued joint, all while using less liquid than a spritz with the spray bottle. You also have more control over which surfaces get hit with the accelerant.

Kicker has a sickly-sweet smell that some people cannot tolerate. You may find you are one of these people. If you are, you can use baking soda as a replacement. After applying your glue, sprinkle some baking soda onto the glue to start a rapid reaction. The glue will cure within moments, but the baking soda adds some mass to the glue. It is not as clean as the liquid accelerant, but you can use CA glue and baking soda to reinforce joints and fill gaps.

To the right is a block of ABS plastic with two examples of cured CA glue. The left side cured with kicker and the right side with baking soda.

When using CA glue, be careful with how much you use. Squeeze-out and stray drops can easily ruin the finish of your prop. The glue dries hard and tends to take on a crusty appearance when it interacts with some paints. If it gets somewhere you do not want it, you may need to sand it smooth and reapply your paint.

The glue can also bleach some plastics, so test it on some scrap filament first before using it on an unpainted model.

> **Warning!**
>
> If you use enough CA glue that you can smell it (or make your eyes water), you have used far too much. Be careful and use only small amounts of CA glue. You can develop an allergy to CA glue with repeated exposure to high volumes of the glue's vapors.

High-Temperature Hot Glue Sticks

A hot glue gun should be in every maker's arsenal. High-temperature hot glue sticks work well on plastic parts and most other materials you might use for your projects.

The glue sticks are thermoplastic, just like your FDM filament, so the glue is sensitive to temperature. If the glue gets too hot or cold, it might not maintain its bond. Don't rely on hot glue for important structural pieces.

This potential impermanence can be an advantage. You can usually peel away hot glue to disassemble your prop if you need to replace a part. Hot glue is an excellent option for quickly assembling something that involves porous materials you just need to stick together–like a 3D print you want to be attached to some cloth or foam.

Hot Glue Painted Black

One potential downside is hot glue has more mass than other adhesives. If you apply a little too much, you can create a noticeable gap between your two parts.

On the other hand, you can use this mass to your advantage by applying it along a seam rather than in-between two pieces–like a weld.

In fact, the careful application of a line of hot glue can look like a weld. You can use it to recreate textures on your props. On the left, I used hot glue on this part of the Ghostbusters proton pack to re-create the look of the epoxy that is gooped around this part on the screen-used prop.

Creating a "Seal" from Purple Hot Glue and a Pin

If using hot glue to make a weld sounds interesting, you might check-out colored hot glue sticks. You can buy hot glue sticks in various colors to help the glue blend into your prop or sell it as another substance, like a melted wax seal.

You can also paint hot glue reasonably easily. I have had great success using flexible two-in-one (paint and primer) paints like Plaid's FX flexible acrylics.

Warning!

Hot glue is–of course–*hot*. Watch the gun's tip while the gun is resting on your work surface and while using it. If you aren't careful, you could burn yourself or deform your model.

You will inevitably get some hot glue on your hands or fingertips. If this happens, rub the glue to quickly cool it. You will simultaneously avoid a severe burn and remove the glue.

There are also other creative uses for hot glue. You can fill cavities in a printed part–like a hollow resin print. The hot glue will add weight to the part and provide structural support.

You can also use it to fill silicone molds for quick reproductions. Here is a collection of greeblies made from colored hot glue sticks. I slowly squeezed the molten glue into silicone molds designed by Ted "EvilTed" Smith.

Hot Glue Squeezed Into Silicone Molds

Hot Tip

While it won't hold anything together, you can buy sticks of colored wax meant to be used with hot glue guns set to a lower temperature. They're often sold as a wedding supply for sealing fancy invitations. You can use them to decorate props and costumes.

Two-Part Epoxies

Two-part epoxies are another option for a fast curing adhesive. They work well with most materials and are usually set within five minutes. Epoxy is a thermoset, so it cannot be melted or return to its liquid state.

Your typical epoxy is packaged in two conjoined tubes with a plunger that pushes out equal amounts of the two parts for mixing. Some epoxies will require you to mix parts A and B from different containers.

Epoxy is self-leveling (gravity will smooth it out) and cures hard, transparent, and glossy. You can selectively apply a thin layer of epoxy to a surface to give it a wet appearance. This is a great way to grant a small area a glossy finish on an otherwise matte prop or model.

You can also dye some epoxies. On the right, I used purple dye with Smooth-On's XTC-3D epoxy and applied it to this moai statue[21] I printed in PLA. The epoxy smoothed over layer lines and gave the figure an amazing glossy surface.

This epoxy is not meant to be used as an adhesive. It has a much longer pot life (it cures slowly) to give you time to brush it onto a model. You can do similar things with 5-minute epoxies if you only need to cover a small area and want it to cure quickly.

If not dyed, the epoxy hardens to a clear, glossy surface.

Dyed and Brushed Epoxy on a Print

One special type of fast-curing epoxy is J-B Weld. This epoxy is a special steel-reinforced epoxy that works wonders with metal. It's one of your best options for bonding metal to your printed parts (and most other materials). The plastic part will break before the epoxy releases its bond.

Unlike other epoxies, J-B Weld mixes into an opaque grey color. It does not appear glossy or look attractive on a prop. Apply only enough to cover the surfaces you want to bond. Use a small knife, bamboo skewer, or similar small object to carefully apply the mixture to your desired surfaces.

Tip

Epoxy will not stick to silicone, so you can use a silicone surface to mix parts A and B. Let the leftover adhesive cure on the silicone so you can see when the epoxy has hardened. The hardened epoxy will peel away.

If you pour silicone molds, pour extra silicone into a bowl or a cup to create silicone discs for mixing adhesives and paints.

Epoxy Putty

Plumber's epoxy is typically sold in a two-tone stick. You cut-off a piece of the putty, knead it until it is a uniform color, and stick it wherever you need to seal a seam. The putty gets into crevices to form a watertight seal and holds onto smooth surfaces.

You probably won't use this stuff to assemble your models, but it does stick to 3D printed parts very well and can add structural integrity to a joint. If you want to fake a metal weld or sculpt a feature on a print's surface, epoxy putty might be a good solution.

I have used this to create weld lines on multiple props, like this ghost trap. The rear box of the ghost trap has a prominent weld line. I recreated it by rolling out a thin snake of epoxy putty, cutting it into small pieces, and pressing the pieces into the plastic in an overlapping fashion.

A "Weld" Molded in Epoxy Putty

You can use some water and sculpting tools to shape the putty. Once cured, the putty is rock hard and can be sanded or chiseled into shape. I used a hobby knife and rotary tool to smooth and shape this weld line.

The putty usually cures relatively quickly (fifteen minutes to an hour, depending on the formula). If that is too quick for your task, companies like Smooth-On sell a two-part epoxy putty made for sculpting.

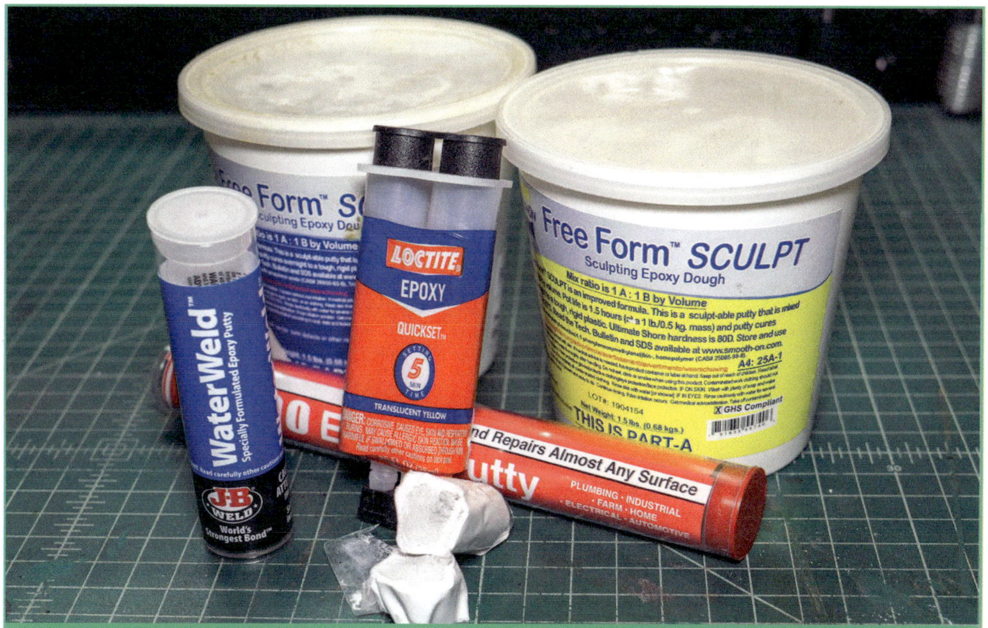

Epoxy Adhesive and Sculpting Dough

If you are working with parts printed with ABS filament, you can bond and repair parts with mixtures of scrap ABS and acetone.

Acetone evaporates quickly unless you have it in a well sealed container, so don't try to stock up on these mixtures. When you need some:

1. Get a scale, some glass jars, and your acetone.
2. Measure out 50mL (50g) of acetone into each of your jars.
3. Take some ABS filament or scrap material (e.g., failed prints or brim and skirt material) and cut it into small pieces.
4. Add the scrap filament to the jars in the following ratios.

> **Warning!**
> Treat these mixtures like pure acetone and wear gloves and a respirator when handling them. Go to a well-ventilated area to apply glue or slurry to a model. Leave the part outside or in the same well-ventilated area as the acetone evaporates.

ABS Juice, 1:10 Ratio

The thinnest mixture, ABS "juice," is a 1:10 mixture of ABS and acetone that creates a solution similar to 2% milk. Mix 4-5g of scrap into the 50mL of acetone, stir, and set it aside to thoroughly dissolve.

You can brush this solution onto a print bed to create a surface on which ABS will stick well. This is an excellent defense against first layer issues if you are experiencing problems prints sticking to your print bed.

Don't apply the juice with a good paintbrush. Use a disposable acid brush. The brush will stiffen with ABS plastic once the acetone evaporates, but you can reuse it for a while. The brush will soften if you use it to stir the juice for a few seconds before using it to apply glue to your print bed.

ABS Glue, 1:5 Ratio

Doubling the amount of ABS to 8-10g creates something akin to Elmer's School Glue. When brushed onto two ABS pieces, the glue acts like a weld-bond glue. The acetone melts (like a weld) the two components together to create one piece. The ABS in the mixture helps you spread the acetone and fill-in small gaps.

This part started as two separate pieces. I brushed ABS glue onto the opposite surfaces and into the seams. Once the acetone dried, I sanded the rough edges, and now the part looks like it was always one piece.

ABS Slurry is Thick and Spreadable

ABS Slurry, 2:5 Ratio

The thickest mixture, the slurry, is thick and gooey with a 2:5 combination of ABS and acetone. Combine ~20g of scrap with 50mL of acetone and wait a while for it to dissolve. You want the slurry to be thick, not chunky.

You can brush this onto an ABS part to fill holes and smooth layer lines. It's great for filling seams after using ABS glue or as a stand-in for body filler.

Hobby knives and metal spudgers work well as application tools. Put a small amount on a flat edge and press it into gaps and holes. Use the bladed edges to evenly spread the goop across the surface of your part.

Keep a Small Collection of Scrap in a Baggie

Glue is convenient, but mechanical connections are often much stronger. These connections stick parts together using screws, rods, slots, pins, clips, nuts, and bolts. I prefer to use mechanical connections for my props wherever possible. The assembly is reversible for easy repairs and access to electronics. Mechanical connections do require much more foresight and planning because you need to plan screw placement and other assembly steps in advance.

The Pack's Hidden Mechanical Attachments

Much of the initial planning phase for my proton pack build involved figuring out how to assemble everything. I wanted it to be easy to repair or replace parts. I used nuts and bolts, wood screws, and a variety of socket head and socket cap machine screws to assemble all of the individual components.

The only glue I used in the build was hot glue. Hot glue holds the plates to the bottom of the pack but can be cleanly removed from the truck bed liner I used to paint the shell.

Clips & Snaps

There are many ways to connect plastic parts together (temporarily or permanently) without glue or screws. You only need to look around your house for inspiration. A snug fit and friction is often enough to secure a part. You can add clips to help keep everything in place.

Look at just about any remote control you have in your house. The battery compartment is probably behind a sliding cover that has one or more clips on it. The clips fit into slots with just enough tension to keep the cover from shaking loose.

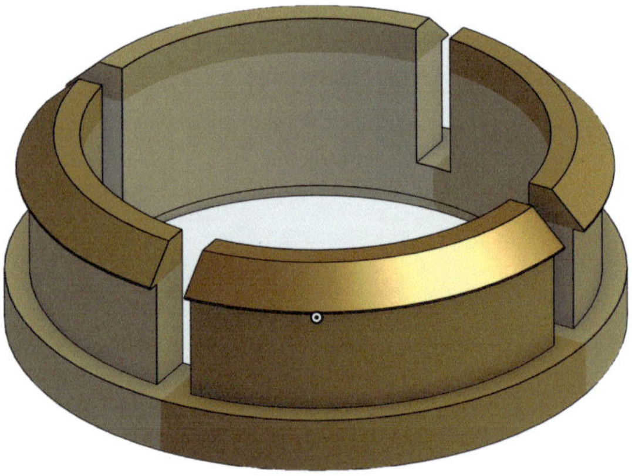

On the other end of the spectrum, there are clips on items like water-proof cases that are meant to connect permanently. I designed this ring with such a clip. The ring fits into the hole on the proton pack, where a hose enters at the bottom. I created the clips to flex enough to go into the hole but spring back into position once inside the pack. It's a permanent connection; removing it would require destroying the clips from the inside.

Note

If you plan on using heat inserts, add extra shells around the holes. The additional shells are insurance against the insert melting through the first two or three shells you might typically use.

ABS Print with Brass Inserts

Using Screws

Screws create strong connections, and exterior screws often look great as part of a prop. The biggest downside to using screws is how easy it is to strip threads. Overtightening a screw can shred the delicate threads the screws cut into the soft plastic. That leads to loose fits, sagging, or screws that just fall out.

Machine screws are some of the easiest to overtighten and strip. They have fine threads intended for metal, but they often look the best for props. Socket head and cap screws are found all over different types of props.

If you don't need a specific screw for appearances, wood screws work best for biting into plastic. These screws are made to sink their deeper threads into soft materials. Consider using them in hidden areas where no one will ever see them.

Wood screws are more likely to split your plastic parts. Make sure you have a hole that is only slightly smaller than the screw. If you've ever hastily driven a screw into a piece of wood, you already know that getting aggressive with a wood screw can quickly lead to disaster.

Cutting threads into plastic works well for many props. For props that you will handle frequently, consider threaded metal inserts for extra durability.

Threaded Inserts

These inserts are metal (usually brass or stainless steel) and add durable threads to holes in softer materials, like wood or thermoplastics. They also look great!

There are two common types of inserts for 3D printed parts: heat inserts and expanding inserts.

Heat inserts require a soldering iron and special tips. The tips fit inside of the inserts and heat the metal until it melts into the plastic. As the plastic cools, it solidifies inside of the grooves and knurling machined into the insert's sides. The result is an insert firmly embedded in the plastic.

Be warned, heat inserts only work for thermoplastics. Thermoset plastics, like resin prints, do not melt.

Expanding inserts work for thermoplastics and thermosets. You press these inserts into a hole without heat or any special tools. The inserts have knurling on the sides that bite and cut into the softer material for a secure fit. They also expand when you insert a screw, pushing the knurling deeper into the surrounding material.

If you accidentally drill a hole that's too big or strip a screw, you might be able to repair it by adding an insert.

M3, M4, and M5 Expanding Inserts by E-Z Lok

Drilling Holes

I try to avoid drilling into 3D printed parts. You are better off pre-planning screw holes during your design phase and using drills for clean-up or slight enlargements. Still, there will be times you just need to drill a new hole into some plastic.

Drilling into plastic–3D printed or not–is stressful (for you and the plastic). Drill bits love to bite into the softer material and wander. It's easy to engage a drill intending to slightly enlarge a hole and suddenly have the bit punch out the other side of your part or cut sideways.

You must move slowly and progressively increase the size of the hole. If you try to move quickly, your larger drill bits will drive erratically and create ragged non-circular holes. Possibly even in a different location than you intended.

Punch-out and chipping are also concerns. Plastic parts can chip or crack when your drill bit cuts into individual layers or exits the other side of a shell. Remember, the sides of a 3D printed part are less than a few millimeters thick and made-up of individual plastic lines that are perpendicular to your drill bit. Regular drill bits can snag layers at odd angles and wreak havoc. You can drill small holes without much issue, but you will have problems as you go beyond ~1/4".

I drilled several holes into this painted test block of red ABS plastic. The initial cut of the 1/8" drill bit created a small paint chip, but the hole in the plastic was clean. Increasing the size to 3/16" and 1/4" created more ragged holes.

Forget any notion of using a Forstner bit for larger holes in a printed part, too. Even smaller Forstner bits are likely to cause cracks or splinters before they cut into the plastic.

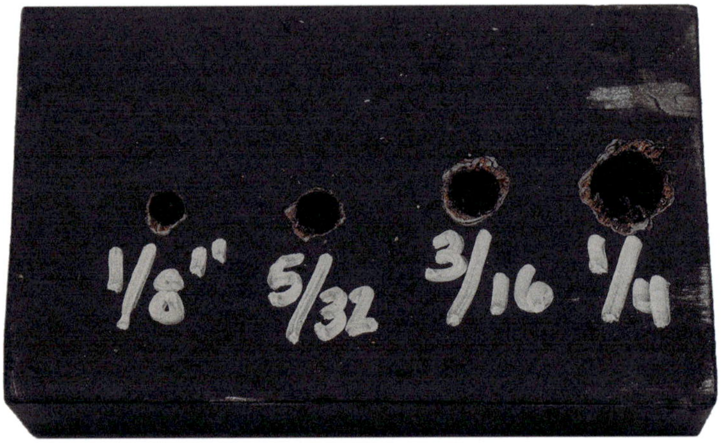

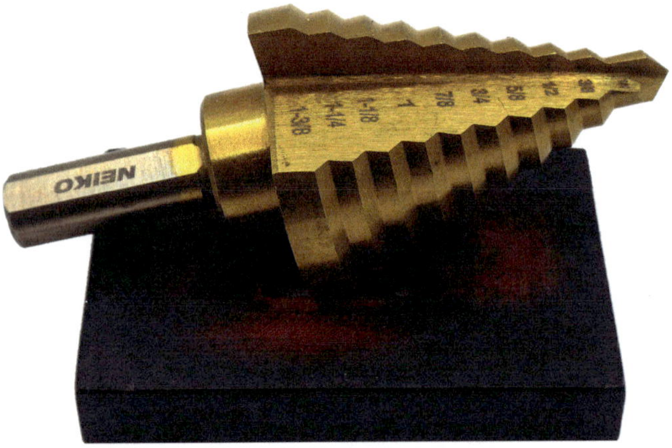

Luckily, there are drill bits made to make all of this much more manageable–step bits. These special drill bits have bladed steps that get progressively wider. Using a step bit is the safest way to cut holes into plastic (printed, vacuum-formed, or injection molded).

Tip

Start out using your step bit in reverse. The bit won't cut into your part but will create an indentation. Check the placement of the indentation. If it's in the right spot, place the bit in the indentation and start drilling slowly. The indentation will keep the bit in the right spot as it cuts the initial hole.

Tapping Threads

If you expect to use screws for assembly, a set of taps will serve you well. You match a tap to the screws you plan to use. The tap drills a properly sized hole and cuts threads so your hardware screws in smoothly. You can purchase taps meant to be rotated by hand and taps for power drills.

You can use a regular drill bit and cut threads with the screw, but a tap offers some big advantages:

- Removes the guesswork out of the drill bit selection
- Ensures the hole is the right size and uniform, so screws don't get stuck.

The second bullet is more impactful than you might expect. If a screw gets stuck and can't go any deeper, it may twist in place and ruin the threads. Drilling the hole with a tap ensures any extra material is cleared, and your screws will rotate smoothly from top to bottom.

Be warned, tapping a hole with too much torque or speed can have a similar effect. The tap will rotate without moving forward (deeper). Work slowly for the best results.

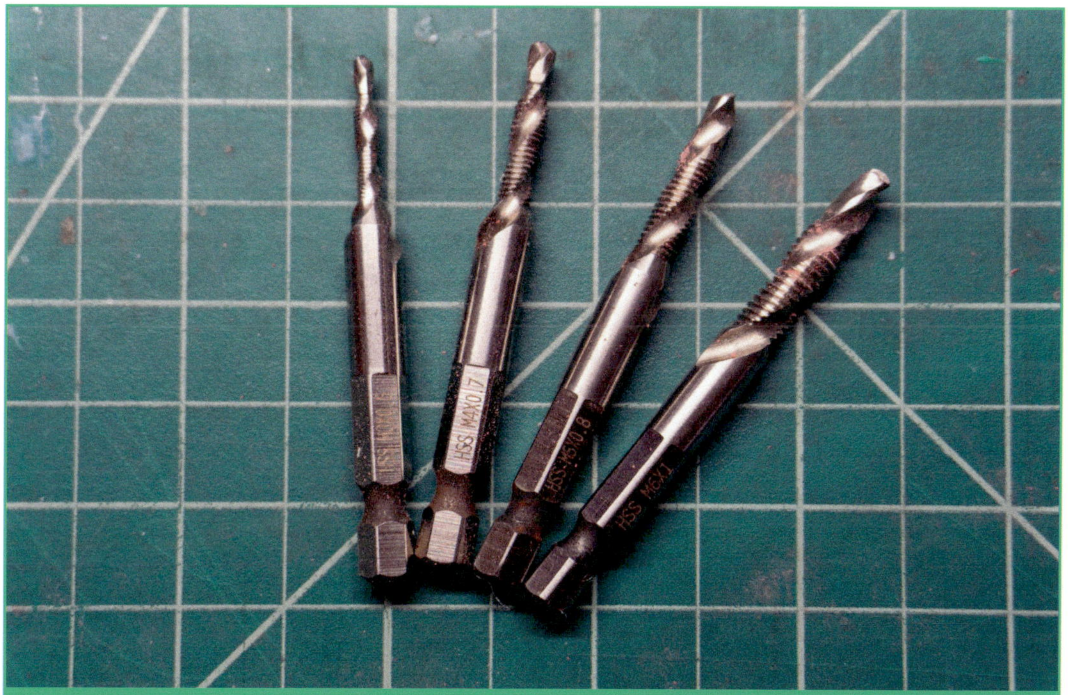

Set of Basic Metric Taps for M3, M4, M5, and M6 Screws

Tip

Sometimes it's better to forego the power drill and use your hand. You will have no trouble rotating a sharp drill bit by hand. This ensures you never accidentally use too much speed or power, and you can be very precise.

Preparing Parts for Painting

The journey towards finishing a part begins with some general clean-up–e.g., remove support material, peel off any brims–and then sanding. There is a lot of sanding.

Plan ahead, so you know what you need to sand. By now, you have test fit your parts and know which surfaces will be visible. You shouldn't expend much effort sanding non-visible surfaces. Clean-up any rough spots that will affect fit and finish and then focus on the areas you want to be pretty. The first goal is removing visible layer lines and blemishes.

Sanding

Sanding requires no special skills, but you need to approach it with the right tools and plan. Sandpaper is your go-to option and easy to obtain. The number one rule of sandpaper is to replace it often. Sandpaper will quickly get clogged and shed its abrasiveness, especially cheap sandpaper. Change-out your sandpaper frequently.

In addition to sandpaper, sanding sponges are useful for curves and irregular surfaces. Emery boards–or nail files–and metal needle files are also excellent options for smoothing and shaping flat surfaces. The rigid boards and files give you nice surfaces and corners in areas where using your fingers and sandpaper might create unwanted curves or depressions caused by uneven pressure. You can make larger versions at home by gluing strips of sandpaper to popsicle sticks, tongue depressors, and paint sticks. You can do the same with wooden dowels to sand curved surfaces.

As you sand a part, the plastic dust will get into crevices and obscure scratches and layer lines, so periodically stop and wipe down the part. As you wipe away the dust, you will begin to see shiny spots. These spots are the low parts of the model that evaded your sanding efforts.

Speaking of dust, wear a dust mask and try to keep your workspace clean with a shop vac. You will quickly kick-up clouds of dust. I briefly sanded this part with 220-grit sandpaper and collected the dust into a sizable pile. Even small jobs generate a lot of dust, and it spreads like it has a life of its own. You do not want it into your lungs, so wear a mask.

A Lot of Nasty Dust

Cover Yourself

Plastic dust will get all over anything within a few feet of your workspace, including you. Wear an apron or shop hoody and keep any drinks or food covered–or *far* away.

I usually begin with 220-grit sandpaper, 150- and 220-grit emery boards, and 180-grit diamond needle files. Don't be too quick to start with these finer grits if your part needs more significant reshaping. An 80- or 100-grit sandpaper will serve you better for shaping rough surfaces, like those that faced support material. Work your way up to the finer grits as needed.

Continue sanding with 220- or 400-grit until the surface is entirely uniform. Once the model has the right shape and smooth surface, move on to 320- and 400-grit sandpaper to remove the 220-grit paper's scratches. Then apply for the first coat of primer. A light coat of primer will highlight any spots that need more sanding.

Fresh Print

220-Grit Sandpaper

320-Grit Sandpaper

400-Grit Sandpaper

I sanded this pokéball with 220-, 320-, and 400-grit sandpaper. You can see it progressively become smoother and more reflective. Still, if you look at the final photo closely, you will see some visible layer lines and low spots that had been hiding behind some packed ABS dust in the other photos.

The imperfections are tiny, but they would become more noticeable with primer. You can always sand more, but you eventually need to decide when the surface is "good enough" and move forward.

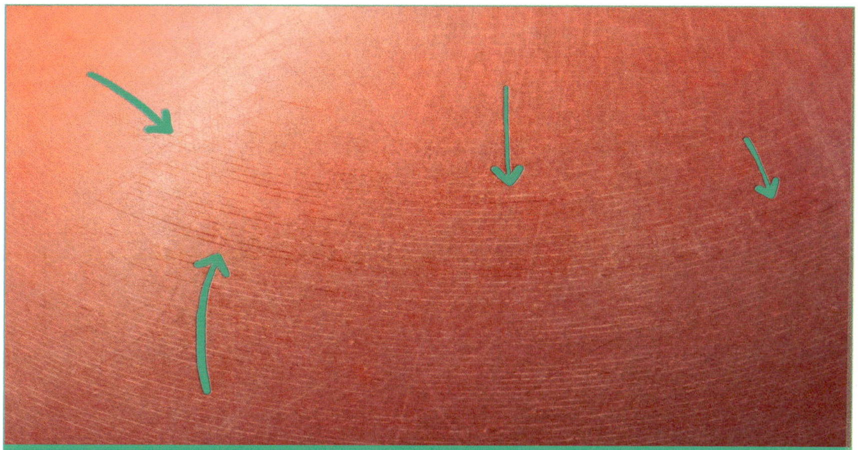

Sneaky Low Spots, Easier to Detect with Primer

Polishing

If your part is already the correct color, you may not need to paint it. Continue sanding and polishing until you are happy with the surface texture. Once you are done, a light pass with a heat gun and some plastic polish will restore the plastic's original color and give it a nice shine.

I finished the red bars on this ghost trap without any paint. I used only 220-, 320-, 400-, 800-, and 1200-grit sandpaper and plastic polish to achieve a glossy finish. I wanted the part to look like lightly scratched, weathered, and worn plastic. A painted surface would not have had the same finish or shown the slight discolorations in the plastic.

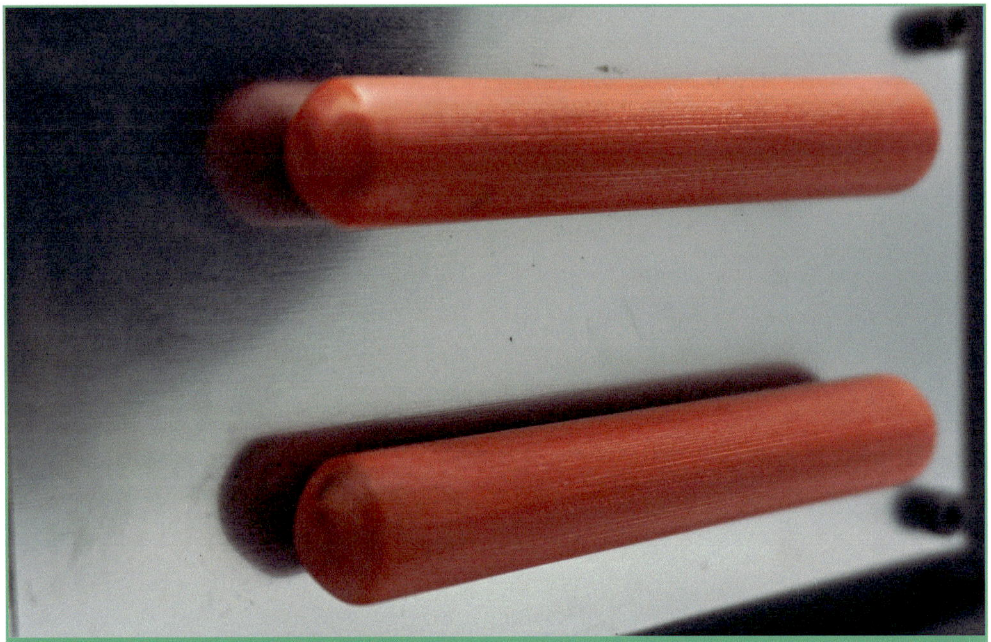

Polished Bare ABS

You may also want to polish models like this before making a mold (more on that later) or when you want a pefect high-gloss or metallic finish.

I followed the same process with these red ABS parts to make them really shine. The visible black lines are faint traces of a light coat of primer. The white is residue from the plastic polish. These parts were silky smooth but some primed areas needed a little more work to remove traces of the deeper lines made by the needle files.

Polished to a High Shine

Body Filler

You may find imperfections that you can't fix with sanding, like extreme stair-stepping on FDM prints or holes in resin prints. It is also common for supports to leave little divots or holes in SLA prints. Avoid sanding too much or too aggressively to fix these issues. You may remove too much material.

You can combine CA glue and baby powder (talcom powder) to create a paste that cures quickly and fills deeper gaps. You can use an accelerant to cure it immediately and create a hard, sandable surface that has a strong bond with your model.

If you have some small gaps or a rough shell (like a surface that faced support material), spot putties from Bondo and 3M are good options.

> **Note**
>
> If you realize you need body filler after you already primed the part, sand away all of the primer in that area for good adhesion. Body filler can also interact poorly with some primers. It strips the paint and creates a gummy mess.

Filling Rough Support Surfaces

These products fill-in problem areas on your printed part. Once dry, you can sand the spots smooth and paint over the filler.

Tubes of this stuff are moderately expensive ($16-25 per 14.5oz tube), but 14.5oz goes a long way. You will only use small amounts at a time to fill in blemishes and imperfections. You may not need to buy a new tube for a long time.

I recommend 3M's Acryl-Green spot putty. The color designates how quickly the putty dries. The Green putty dries the fastest (within 15 minutes), making it a great option for filling imperfections in resin and plastics.

Apply filler in small amounts. If you slather it or try to fill a deep hole, the putty will never dry. 3M and Bondo recommend keeping applications thinner than 1.5mm (or 1/16"). Apply multiple layers if you need to fill a deeper scratch. If you need more than a layer or two, CA glue paste might be a better option.

Apply the putty using a metal spudger or a small knife. Press the putty into the blemish firmly and slide the blade away.

You may encounter an issue using spot putty on smooth surfaces (like resin and freshly extruded plastic) or in shallow scratches. The not-so-obvious answer may be making the hole wider or deeper.

Use a rotary tool with a small cutting bit, a needle file, or some sandpaper to rough-up the surface and make the scratch or hole slightly bigger or deeper. This will give the surface some teeth and depth to grip the putty.

Gaps Filled with 3M Acryl Green

Once it's dry, sand lightly to avoid removing all of your putty. The putty gives way to sandpaper faster than the plastic. It is easy to sand too aggressively and removes more putty than you intended. The above photo had a small but deep groove left behind by the removal of support material. I filled it with Acryl Green and sanded it slowly until I had removed any filler on the top layer. This left filler inside of the groove and a few other small gaps in the surface.

Applying Primer

Do not skip priming your parts before painting, even if you have a two-in-one (primer and paint) rattle can. A can of primer is your magic wand. Even after all of the sanding you have done to get here, your first primer coat will usually reveal areas that need more love.

Cold Rattle Can

To maximize the primer's effect and get the best finish possible, warm-up your rattle cans before using them. A rattle can that has been sitting on a shelf will spray unevenly and offer poor coverage.

To the left, a "cold" rattle created this blotchy surface. The primer sprayed unevenly, so one area got a thicker coating than the rest of the surface.

Note

Don't soak your rattle cans in ***hot*** water. You want water you can comfortably touch–like a warm bath. The goal is to lightly warm-up the can, not heat it. Don't ever heat compressed cans of anything.

Warm Rattle Can

After sitting in warm water for two minutes, that same can of primer sprayed a fine mist; this greatly improved coverage and surface quality.

Primer Highlighting Layers & Stairs

I warmed-up a rattle can and sprayed a light coat of primer on this print. The primer sunk into some crevices and made them much easier to see.

I like to begin with one or two light coats of automotive filler primer. This primer prepares the surfaces for paint and builds up in crevices to help hide light scratches and layer lines. I keep several colors and types of primers on hand for this phase of painting.

Take it easy with filler primer: heavy coats will fill-in or obscure surface details.

Let the first coat of primer dry (if you can, let it sit overnight or spend an afternoon in the sun) and then start another round of sanding. Hit the problem areas the hardest, but do not forget to smooth other primed surfaces. Some primers have a texture you (probably) do not want on the surface of your final piece.

You may sand away most of this first layer of primer, and that's OK. This first layer is really intended to help you see what primer is left behind after sanding. The leftover primer identifies the part's high and low spots. Also, you should use light coats of primer. If your first primer coat is still there after sanding, you may have applied too much.

You want to cover the bare plastic with a smooth finish without adding more thickness than necessary.

While you do this, take it slow while sanding the filler primer. It will quickly build-up on your sandpaper, so you will have to pause and re-fold or swap out your paper. If you sand too aggressively, this primer will build-up on your part as a small mound. If that happens, you will have to remove it with a knife or scraper and smooth over that area.

Once you are happy with the part, apply two or three light coats of a different color primer. It does not have to be a different color, but I recommend alternating colors for this phase. Let the primer dry and examine the part for layer lines and unwanted blemishes.

Sand the part with some 220-grit sandpaper and pay attention to the colors to identify high spots. You will sand through the primer on the high spots and reveal the previous primer's color (or bare plastic).

I started with a grey filler primer for this part on the right and then switched to a charcoal grey. At the end of this third round of sanding, I could still see some problem areas. However, I planned to paint the part a matte black and would attach several other parts to it before the end. This finish was more than good enough.

If you want a metallic or glossy finish, keep sanding the surface smooth. Then, apply another primer coat and move on to wet sanding with finer grits (400-grit and higher).

Red ABS Seen Through Two Colors of Primer

Note

Notice how I did not sand the top of this example print (bottom of the image) very much. This section was covered by another part, so would ever see it.

Wet Sanding

In this phase you put in the work for that silky mooth finish. For large parts, I keep two colors (charcoal and flat red) of regular primer on hand to contrast the typical light grey of the automotive filler primer. Standard primer goes on smoother and wet sands more easily. Some filler primers contain talc (hydrous magnesium silicate) that wants to absorb moisture. Wet sanding a thick coat of filler primer can create a gummy mess.

Primer for All Occasions

I prefer to use my airbrush for smaller or more delicate parts. Alclad's "micro filler" airbrush primer works well with 3D printed models. This primer is sold as a primer for pre-packaged model kit parts that may have light scratches, so I prepare the parts by sanding and polishing them ahead of time.

Per Alclad's instructions, you apply the primer in a few light to medium coats to fill small scratches and polish it with a 3000-grit sanding pad.

Whether you use a rattle can or an airbrush, apply a few light coats of primer and let them dry. To begin wet sanding, fetch a cup of warm water with a few drops of dish soap. Let your sandpaper soak in the cup and then gently sand the primed part using circular motions.

The water will prevent the primer from clogging the finer grit sandpaper, and the dish soap acts as a light lubricant. Work your way up to the highest grit you want to use. For extra glossy or metallic finishes, progress from your 400-grit through 1200- or 2000-grit. Once you're done, wipe down and polish the part with a soft cloth. It should be shiny and reflective.

Note

Put in some elbow grease and really buff the surfaces with a cloth. If your primed surface is smooth, it will shine like injected molded plastic. That shine is satisfying and helps you detect any remaining scratches or blemishes.

Don't worry if this process removes some of that last coat of primer. You will naturally remove more primer on edges, high spots, and other areas with the primer's thinner layers. If you have large bare areas where you sanded away all of the primer, apply more primer before another sanding round.

I sanded and polished the parts on the opposite page before applying a coat of charcoal grey primer. Wet sanding with 800-grit sandpaper removed most of the primer from one of the parts. There was nothing wrong with this; the part still got a shiny smooth surface.

Re-apply light coats of primer as needed and continue until you are satisified with the surface. As the part begins to shine, you may notice small imperfections that you missed earlier. These might be easy to fill with one or two additional coats of primer, but you want to be careful not to add so much paint that it affects how your parts fit together. After all, this is just primer; there's more painting yet to come.

If the imperfection is really noticeable (or you demand perfection), your best bet is returning to your 220-grit sandpaper. If you have kept your primer coats light, it won't be too difficult to fix the problem and re-finish that area. If you laid thicker coats, sanding the primer will be messier and there will be a noticeable difference in the thickness of the primer.

Tip

Changing your primer color can make it easier for your eyes to perceive scratches and layer lines. I start with grey and move towards darker colors. For me, it is easier to see obvious problems in lighter colors. Progressively darker colors help me find the smaller imperfections. Experiment and see what works for you.

I sanded and polished the part on the left to a high shine before I primed it with Alclad's micro filler primer and buffed it with a wet 3,200-grit micro-mesh pad. I primed the parts on the right with a rattle can and wet sanded with 800-, 1,200-, and 2,000-grit sandpaper and the 3,200-grit pad.

Airbrushed Micro Filler & Pad

Rattle Can & Sandpaper

The part on the left was smooth and lost little to no primer during the buffing phase. Meanwhile, the pieces on the right had more edges and rougher surfaces, so the sandpaper removed a lot more primer.

Once you are satisfied with the look and feel of your primed parts, it's time to move on to painting. For glossy and metallic finishes, you want the surface smooth and shiny to the point of being reflective.

STOP!

Now is an excellent time to test fit your parts once more. You just added some thickness to all of your parts. If you have pieces that need to fit inside of another piece or move, the layers of paint might make those fits much tighter than they were previously. Also, check any inset screw heads to make sure they don't fit so close they scrape your paint.

Better to learn this now before you put more time and materials into a finished paint job. The last thing you want is one part to fit so tightly it strips the paint from another part during assembly.

Painting Your Models

How you paint your model is entirely up to you. Since 3D prints are plastic, you can use just about any standard type of paint. Painting techniques are outside of this book's scope, but this section walks through a few basic concepts and considerations. The ***Additional Resources*** section has a list of books that dive deep into specific painting techniques and mediums.

Choosing Your Paints

As mentioned above, a large variety of paints work well with plastics. You can use all of these with your models:

- Acrylics
- Lacquers
- Enamels
- Oils

I tend to use acrylics and lacquers for most of my painting work. Lacquers are great because they dry within minutes, especially when applied lightly with an airbrush. The paints with some of the most convincing metallic finishes are lacquers. Acrylics are available in many colors and hues and easily mixed to create new colors.

You don't need to worry too much about a paint finish when you pick your colors. If you have the perfect red in satin but need a glossy finish, use the satin. The final finish will be determined by your topcoat. Your base layer's finish does matter when you want an extra glossy or metallic finish (more on that later).

Once you have your paints picked-out, take some time to perform a paint test. Unless you are only using paints you are very familiar with or are all part of a "paint system" and meant to be combined, your paints may interact in unexpected ways.

Print a small block or find a failed print and then sand it and apply some primer. Get this piece ready for painting, just like your final pieces. Then paint this test piece like you plan to paint your model. Continue the paint test all the way through applying a top coat and any additional washes you have planned.

Note

If you are painting something thin (or TPU) that will flex, look for flexible paints–like Plaid's FX flexible acrylic paint.

As you work, observe any changes to the paint finish. For example, oil paint washes are fantastic for adding grime and weathering props to appear metallic and mechanical. The oil paints tend to stick to surfaces, slightly discoloring the base layers and adding a greasy shine. The oils can stick too well to Alclad and Testers' metallic lacquers and strip the lacquer when you wipe away the wash. You can avoid this by sealing the lacquer with a top coat before applying the wash.

You may discover similar interactions during your test, so test everything before doing it to your real prop.

To continue the lacquer example, I painted this amulet[22] from *Indiana Jones and the Fate of Atlantis* using Alclad lacquers. Knowing I intended to weather it with an oil paint wash, I sealed the metallic lacquer with Alclad's Aqua Gloss top coat.

Tip

Take the advice from every bottle of household cleaner: test everything in a discreet or hidden location first–or on a test print.

Later, to make the amulet look like it had been at the bottom of the ocean in Atlantis, I wanted to add some grit. I decided to selectively apply light coats of a paint from Sculpt Noveau that contained fine iron powder. I spritzed the wet iron paint with the manufacturer's Tiffany Green patina spray to cause the iron to actually rust.

You spray the paint with the patina solution while it's still wet and then let it sit for a couple of hours. When I came back, I found a surprise waiting for me. The spray had interacted with the Aqua Gloss and given the piece a very different look than I had intended.

This Interaction Was Completely Random–and Really Neat

I had only applied the paint to a few small areas. Those spots had the expected rust and grit, but the patina had also discolored the top coat. The amulet had a whole new look created by run-off and random chance when I sprayed the liquid. Fortunately, I loved the result. There was more, originally, but I lightly rubbed some of it away with a 0000 steel wool and re-sealed everything with a final coat of Aqua Gloss. This could have been a disaster. Instead, it's something I may repeat in the future to reproduce this effect.

Painting Setup

Whether you are using a brush, a rattle can, or an airbrush, you will need to set up a place to paint your parts. When picking a spot, your first consideration should be temperature. The ambient temperature affects your paints and the surfaces of your model. You will get the best results if everything is at room temperature.

You also need some way to let your parts dry in a warm place, away from pet hair and dust, and with the wet surfaces kept away from anything that might touch them and mar the finish. That last point can be difficult if you have small or oddly shaped pieces.

You can address these concerns by using screws, dowels, tongue depressor, and skewers. If your part has screw holes, thread a screw (one you don't mind painting), and use it to hang your part from some string or wire. You can evenly cover the whole piece with paint and then hang it somewhere warm to dry.

If the part has no screw holes but has non-threaded holes, you can place it on a bamboo skewer and stick it into a woodblock or some foam. You can also thread some stiff wire through a hole to use as a handle and a hanger.

If your part has no holes but has a surface that will be hidden after assembly, you can drill a small hole into that surface. Then insert a stiff wire or a screw to use as a temporary handle while painting.

If you can't drill into that hidden surface–or don't want to–use some double-sided tape to stick it to a tongue depressor or paint stick. These sticks are excellent handles. Just make sure the part is firmly attached to the tape!

That's What I Call a Stake Out

I used some screws and bamboo skewers to hold these parts while I painted them. This especially important when you're working with a glossy finish and can't have anything touching the part while it dries.

If you need a high-gloss or metallic finish, it will be worth your time to explore the resources mentioned in the ***Additional Resources*** section. Creating a convincing metallic finish takes practice and patience. For a high-shine reflective surface, you need at least one glossy base coat.

Before you proceed, you need to review your primed and sanded parts with a critical eye. Glossy paint will highlight every defect on your surfaces, and layer lines don't look great under glossy paint. I painted a spare block of ABS to demonstrate this.

I roughly sanded the right side with 220-grit sandpaper and kept the left side untouched. The top layer (bottom section of each example) looked passable, but the layer lines on the sides looked terrible. The finish looked no better with the metallic lacquer. Brighter and glossier paint further call attention to scratches and layer lines.

In certain lighting conditions, the sanded surface actually looks worse because of the swirls and scratches created by the 220-grit paper.

Rough Surfaces with Glossy Paint

You can use strategically use scratches for weathering. Before applying your glossy base coat, take some steel wool or a Scotch-Brite scouring pad and rub the primed surfaces you want to look scratched or worn. You control the direction and depth of the scratches. Even lighter abrasions will show through. I used this technique on the top of this block to add some light scratches to the exposed top surface where it would naturally collect scratches.

Weathered with a Scotch-Brite Pad Prior to Painting

Metallic paints are meant to be applied in thin layers to avoid mottling and uneven finishes. They need a base coat to boost their shine and provide the proper color foundation. Depending on the color you're after, you may need to layer some paints for the optimal effect.

As mentioned, you often need a glossy black base coat to prepare for a reflective topcoat, but always read the paint manufacturer's recommendations to check for any special considerations.

For example, Alclad and Testers recommend a black base coat for many of their metallics. On the right, this bottle of Alclad II Gunmetal lacquer recommends applying the paint over white, grey, or black primer. Gunmetal, a dark metallic that's nearly black, does not need a glossy coat. A gloss coat wouldn't hurt the finish, but it won't do much to help it either.

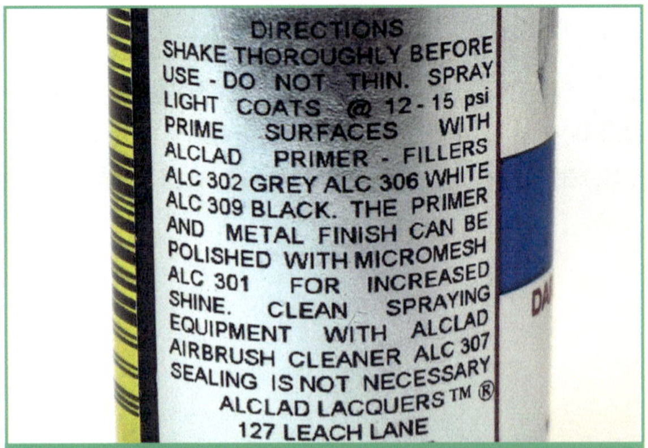

Read the Directions

However, specific colors and paints look their best with silver or gold base coats. "Candy" metallics (think classic candy apple red cars) are transparent and pick-up much of their tone and shine from the base coat. Classic candy apple red relies on a gold base coat to give it a warmer tone.

To achieve your perfect finish, you might need to apply a glossy black base coat, a metallic gold or silver coat, and then your final coat. Applying that first glossy coat can be intimidating.

This bottle of Alclad II Hotmetal Red lacquer is transparent and must be applied over an opaque metallic lacquer. That means you must apply a glossy base coat, a metallic lacquer, and then apply this lacquer.

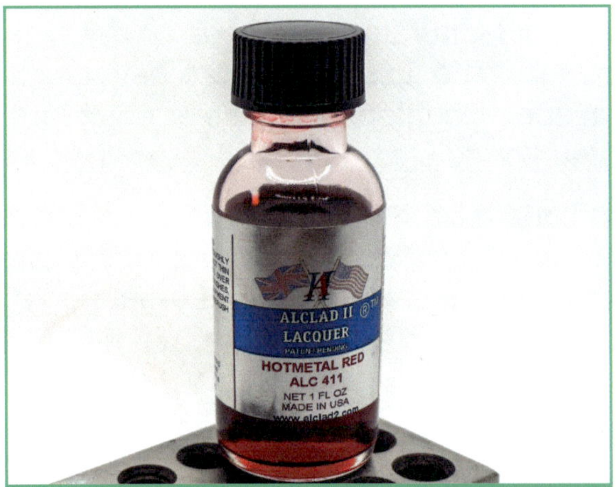

Transluscent Lacquer

For this pokéball, I applied a base coat of glossy black, a second layer of gold lacquer for a warm tone, and finished it with a metallic red.

Metallic Red with a Warmer Tone on a Printed Part

The key to a good coat of gloss is being brave. Start with an initial "dusting," a light coat of the glossy paint. Think about any crevices or low spots that might have been more difficult to prime and hit those areas with light passes of the glossy paint. The paint won't look glossy yet; you've only dusted a base coat of color.

Now it's time to be bold. Make sure you have enough paint to cover your part without stopping and go for it. You want to try to cover as much of the part as possible without stopping. Try to move and swivel your part smoothly, so you get even coverage without hitting any painted areas with overspray. The goal is a surface that looks wet and reflective. Overspray will sprinkle areas with dust and may harm the finish. This step is deceptively tricky because repeatedly covering areas will lead to sagging paint or fisheyes.

Here is a part as it moves through these two stages. I dusted the part, let it sit for a few minutes, and then sprayed a wet coat.

Dust Coat vs. Final Glossy Coat

Here is the same part (after sitting overnight to let the glossy coat fully cure) with just the smallest spritz of Alclad II Steel lacquer and then a full coat. It's like a magic trick.

Light Spray of Metallic and Finished Part

Weathering

Weathering is one of the most fun phases of finishing a new prop. Weathering can mean using paint washes to add dirt and grime or adding some wear and tear via other means. A paint wash or dry brushing does wonders for highlighting little details in low and high spots, respectively.

Products like Rub n' Buff, a wax with high amounts of metallic pigments, work well with 3D printed models. You can use it over paint or apply it directly to the plastic. You don't even need to do any sanding. Rub the wax into layer lines or lightly brush it over surfaces for different effects.

I lightly applied some silver Rub n' Buff to this "Chibi Grim" model[23] [24] printed in a glittery black PLA.

Unfinished Black PLA with Rub n' Buff

For washes, you can use most paints–as discussed in the previous section–and specialty washes from companies that make paint systems for models and miniatures–e.g., Tamiya and Citadel.

> **Note**
>
> Many of the stronger paint cleaners and some thinners contain acetone. If you'll be spraying anything on your model for a paint wash, use water or IPA.

In general, you can do anything you want to do for weathering as long as it does not involve high temperatures or solvents that will melt your model. That said, 3D printing does create opportunities to use these elements for great effects.

It's hard to beat the look and feel of real damage. Thermoplastics are soft enough to be scratched. The previous section mentioned abrading primed surfaces with a Scotch-Brite pad for scratches, but you can go even further. 3D prints deform easily and can even be melted relatively safely. You can get creative with blades, heat guns, soldering irons, and woodburning tools.

As alluded to earlier, you can use a heat gun to warp finished FDM 3D prints. Applying heat to a print will have different results depending on the print's geometry and infill. If the part has infill, the outer shell will begin to collapse into the infill, and the pattern will become visible on the print's surface. The careful heat application will keep the outer layers intact while letting them droop between gaps in the infill. The infill pattern will appear to be embossed on the print's surface.

Heating a Test Print Over a Heat Gun

I printed the Nuka Cola caps from an earlier section with real-world dimensions–i.e., close to .22mm thick. I could quickly heat the caps with a heat gun, and they would cool just as rapidly.

I heated them, placed them on an old glass bottle, and used my Leatherman multi-tool to dent, crease, and flare the bottle caps. The prints were thin and fragile, so they became even more delicate and pliable after only a few seconds over the heat gun.

Final Versions of the Deformed Nuka Cola Caps

Once heated, the caps naturally drooped and became easy to mold. Each cap came out unique, looking like it had its own story to tell. I reinforced inside of the thin–now deformed–ABS with CA glue and baby powder and baking soda. Once painted, the gritty CA glue filling looked like dirt and rust.

Note

The caps in the *Fallout* universe seem to be twist-off, but this technique produced unique caps that looked like I had taken them off of a super mutant!

Melting a Failed Print with a Soldering Iron

You can crank up the heat–literally–and use heated metal to melt parts of an FDM print. The obvious candidates for this are soldering irons and woodburning tools, but you could heat a hobby knife blade, wire, rods, and other things to brand, gouge, and deform the print.

Creating holes and gouges with heat produces fantastic textures that you would not get otherwise. If you're up for it, a soldering iron can help you recreate the sort of battle damage you might associate with Sci-Fi blaster fire or an energy blast.

Melting a Failed Print with a Soldering Iron

You can also use this technique to create welds. I do not recommend this for general purpose assembly, but it can add some excellent texture and help reinforce a joint.

I ran a soldering iron over the joint where these two pieces of PLA met. The heat fused the parts together. Then, I took a bit of filament and used it to reinforce the joint by pressing the filament into the joint with the soldering iron. It created a nice texture.

Warning!

Be careful when using heat to modify your prints. Not only do you want to avoid burning yourself, some plastics and paints may release nasty fumes when heated. Consult the manufacturer's safety sheet and wear a respirator.

Additional Resources

Other makers and artists have published some fantastic books and videos covering painting and weathering using rattle cans, airbrushes, and washes. Like Steven Meissner, some have written helpful guides focused on specific topics like how to properly use body filler.

Here are just a few resources you might check-out if you want to learn more about other finishing techniques:

- *The Book of Cosplay Painting with Brushes and Acrylics*[25] and *Advanced Cosplay Painting: – Airbrush and Weathering*[26] by Svetlana Quindt
- *Painting & Weathering for Props & Replicas*[27] by Harrison Krix
- *Professional Paint Jobs*[28] by William Morgan
- *A Robot's Guide to Sanding*[29] and *A Robot's Guide to Bondo*[30] by Steven Meissner
- *How to Print a 3D Gun Prop from Destiny, a* YouTube video series by Bill Doran[31]

Pokéball

Pokémon

Model by Scott Clarke[32]

Printed in ABS, painted with Vallejo acrylic paints, and assembled with an alluminum RGB LED button

Reproducing Parts with Molds

Finishing a single part can take a while. If you need several copies of the same part, that's more print time and more sanding. There is an alternative: finishing one part and then making a silicone rubber mold.

With a mold, you can rapidly reproduce your fully finished part within a few minutes. Silicone captures your model's most delicate details and surface qualities. A mold of a highly polished part or one coated in a glossy paint will produce shiny parts.

Invest your time and energy into perfecting one part, create a mold, and then cast as many reproductions as needed.

Entire books can be (and have been) written on this subject. This book will cover some necessary information to help you get started. The end of this section has some suggestions for additional resources.

Equipment & Material

You may already have the necessary equipment you need to make a mold. Here are some of the supplies you will need:

- Foamcore sheets
- Scrap wood
- Hot glue
- Plastilina clay
- LEGO bricks

You can use materials like these to build a mold box, a temporary box built around your master (your original part) that creates a silicone boundary. You can hot glue bits of foamcore board to a scrap piece of wood, make a box out of LEGO bricks, or come up with some other way to construct a container. You can also just use tupperware or a jar if you have some the right shape and size. For round molds, you can repurpose things like used paper towel rolls or poster tubes.

The plastilina clay is inexpensive (you can get a couple pounds of it for $5-10), intended to be reused, and has many uses for mold making. Use it to make a mold box, plug holes in the master, or prepare the master for a two-part mold.

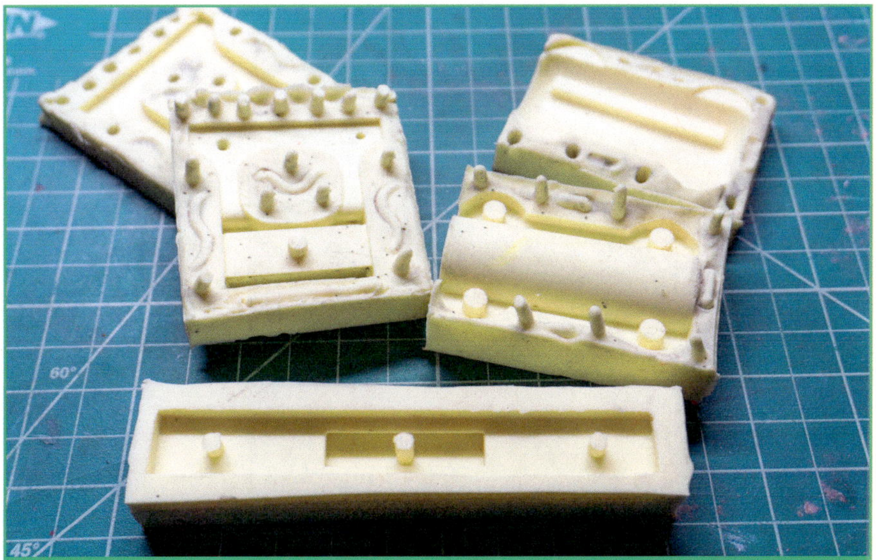

Molds for a DL-44 Scope

These are some of my very first molds. I made several mistakes, but they're serviceable molds, and I learned much from the experience. Trying your hand at making a mold is relatively inexpensive. A "trial" pack (one quart) of Mold Max silicone costs around $30 and has enough for one big mold or several small molds.

Casting Options

The most common material for casting parts is urethane resin. This resin is similar to the photopolymer resin used for SLA prints. Both have a shore hardness rating in the same range (about 70-82D), sand easily, and are reasonably durable. Urethane resin comes in opaque and translucent varieties, and you can dye it to match just about any color.

Even better, there are different types of fillers you can purchase to mix into the resin. Smooth-On, one of the largest manufacturers of silicone rubber and casting supplies, creates fillers like their *Glow Worm* powder that makes your casts glow in the dark. Their *URE-FIL* series includes fillers that give casted parts the appearance of wood or stone, make casts glow under UV light, or add weight for heftier parts.

I used a translucent resin (Smooth-On's Smooth-Cast 326) and an orange tint to create replicas of the iconic amber grips for Deckard's blaster from *Blade Runner*. I could have used a transparent orange resin with an SLA printer, but a mold let me experiment with tints to get the amber color I wanted and a perfectly clear cast (this required a pressure pot).

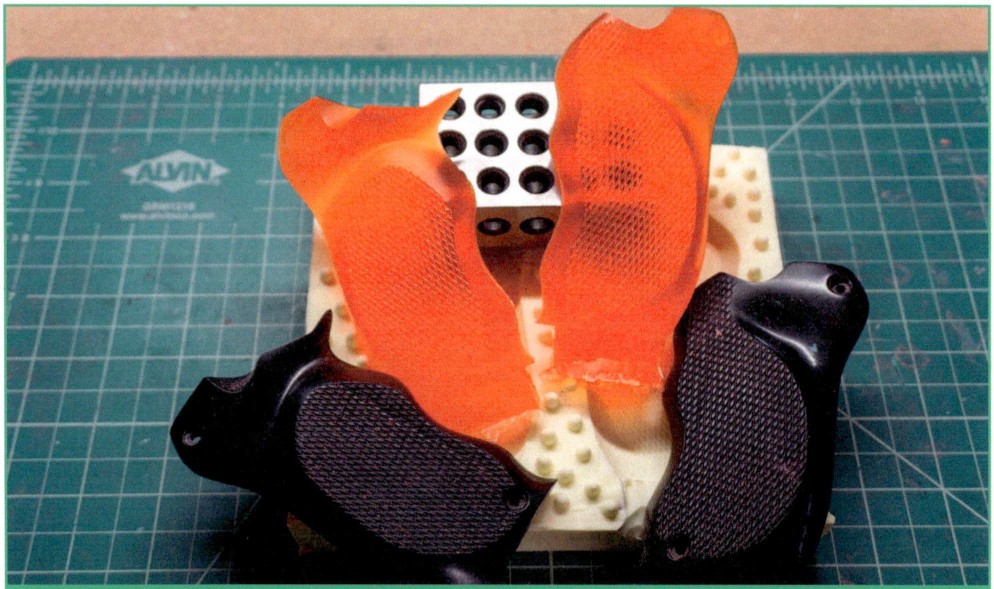

Original Prints and the Molded Replicas

You can buy transparent photopolymer resin and even dye it yourself to make adjustments, but a mold gives you finer control over the final piece's color and opacity.

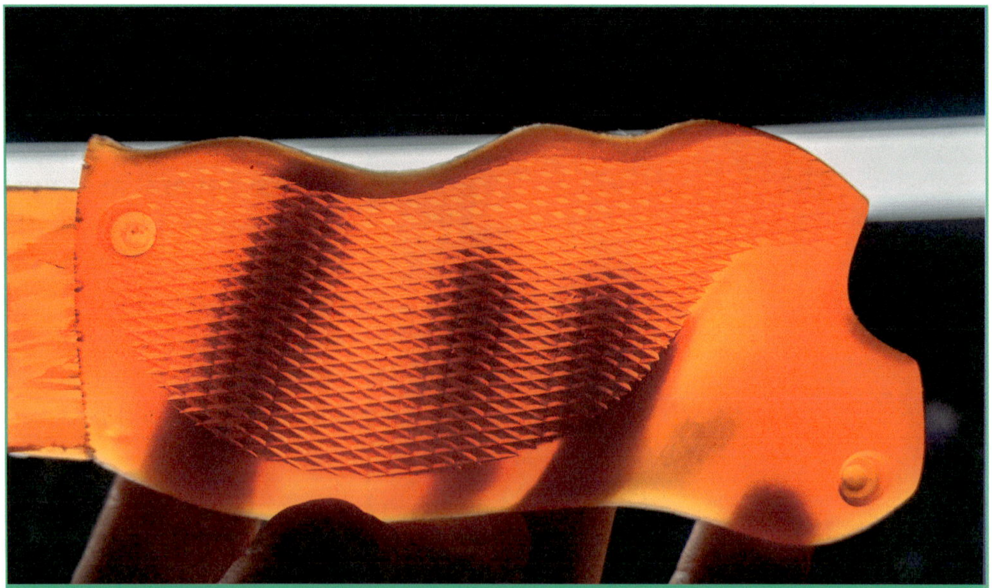
Crystal Clear and Transluscent Replica

The Original PLA Part

Fine Aluminum Metal Powder

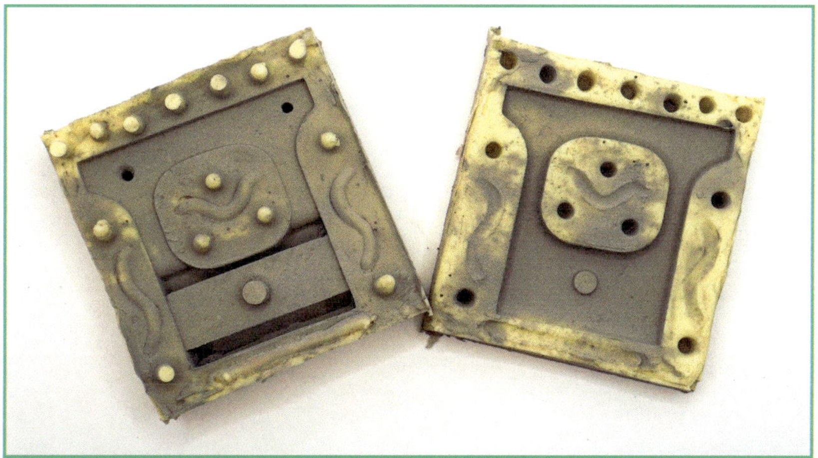

Mold Dusted with Metal Powder

One of the neatest things you can do with molds is cold casting. This technique involves dusting the surface of your mold with fine metal powder before you pour the resin. The resin picks up the powder, and the final part has a thin coating of actual metal. You can then buff and polish this coating to make props look like solid metal pieces without any paint.

I cold cast the scope mount for this DL-44 blaster to get a real metal finish.

A Real Aluminum Finish

Resin is just one option available to you. With a mold, you can make castings with many other materials. You can 3D print grips for a blaster and then cast them in rubber or foam. If you have a material you can inject or pour that will solidify as it cools/reacts/is exposed to air, you can probably cast a prop with it. That includes making edible props out of something like chocolate!

For lightweight and convention-friendly pieces, there are foams you can use with your molds. There is EVA foam clay, a super-moldable material that air dries to the consistency of EVA foam sheets, that works well pressed into molds. Smooth-On also has a *FOAM-iT!* product line of expanding foams that range from rock hard to squishy (like a NERF dart).

My favorite material is pewter. Smooth-On makes a high-temperature silicone rubber, Mold Max 60, that can withstand liquid pewter's heat. The resulting cast has a nice heft to it and can, of course, be finished and weathered like metal. The parts have a lovely shine after post-processing them with 000 or 0000 steel wool and metal polish. You can also use a chemical like JAX Pewter Black to quickly age and blacken the pewter's surface. Rub the aged piece with steel wool to remove the blackened surface on the high spots to create a fantastic contrast effect.

We'll take a closer look at this process in the ***Bringing it All Together*** section!

Parts of a Blade Runner Blaster Cast in Pewter Alongside Their Resin Versions

Additional Resources

If casting has captured your interest, check-out these resources:

- *Cast Like Magic*[33] by Folkenstal
- *Molding & Casting 101*[34], a YouTube video series by Bill Doran

PRINTING
PROPS
2020
A
MAKER'S
JOURNEY
2020
A
MAKER'S
JOURNEY
2020

Custom Challenge Coin Collection

Model by the author

Printed in resin and then cast in pewter, foam, urethane resin, and hot glue
Finished with JAX Pewter Black and a variety of acrylic washes and waxes

Bringing it All Together

This chapter walks through all of the techniques discussed earlier to create something new. Instead of a prop, I chose a classic for makers and aspiring 3D designers alike, a maker coin. The techniques demonstrated in this section easily translate to models for props and cosplay–like creating metal coins and currencies from movies and games.

Maker coins are similar to military challenge coins. They are personalized coins that can function as a business card or a quick way to test out new filaments. These coins are also great starter projects to learn a new 3D design program.

The project goes beyond basic 3D design and printing and explores more advanced techniques, including making molds and casting metal reproductions.

For this book, the coin couldn't be just a filament swatch. The final product is a solid metal coin. To get to that, we'll walk through designing a coin, printing it, making a mold, and casting a new coin in pewter. As a bonus, we will also step through designing a 3D printed case that transforms into a display.

The following sections highlight a few parts of the modeling process. To dive deeper, check out the Fusion 360 files on *printingprops.com*.

Let's get started!

Note

Angus Deveson of Maker's Muse created a video showing how to create his maker coin in Fusion 360[35].

Making a Maker Coin: The Coin

For the coin, I sketched a few ideas and notes on my tablet in a drawing app. As you may have guessed from this book's earlier examples, I have a great fondness for sci-fi and adventure films from the '80s. As a kid, I repeatedly re-watched *Ghostbusters*, *Indiana Jones*, *Star Wars*, and *Blade Runner*, so I naturally gravitated towards replicating props from those films.

To celebrate that, I set out to design a coin that incorporated elements of those inspirations. The design I came up with included a braided edge that looked like an adventurer's rope and would feel nice in your hand.

The map is reminiscent of the charts in the *Indiana Jones* films that showed a plane traveling with dotted lines. I swapped out the plane for a silhouette of the Millennium Falcon for a little *Star Wars*. The Falcon's path shows it flies to the X and ends up back where it started, much like every maker's journey. We begin in one place, reach our destination, learn something new, and finally return to the start to do it all over again. I imagined a 3D silhouette of a ghost trap sucking in the book's title like a ghost for the second face.

I chose to use an SLA printer to produce one solid coin. Using SLA also meant I could get very detailed, not have to print the coin in halves, or do much sanding.

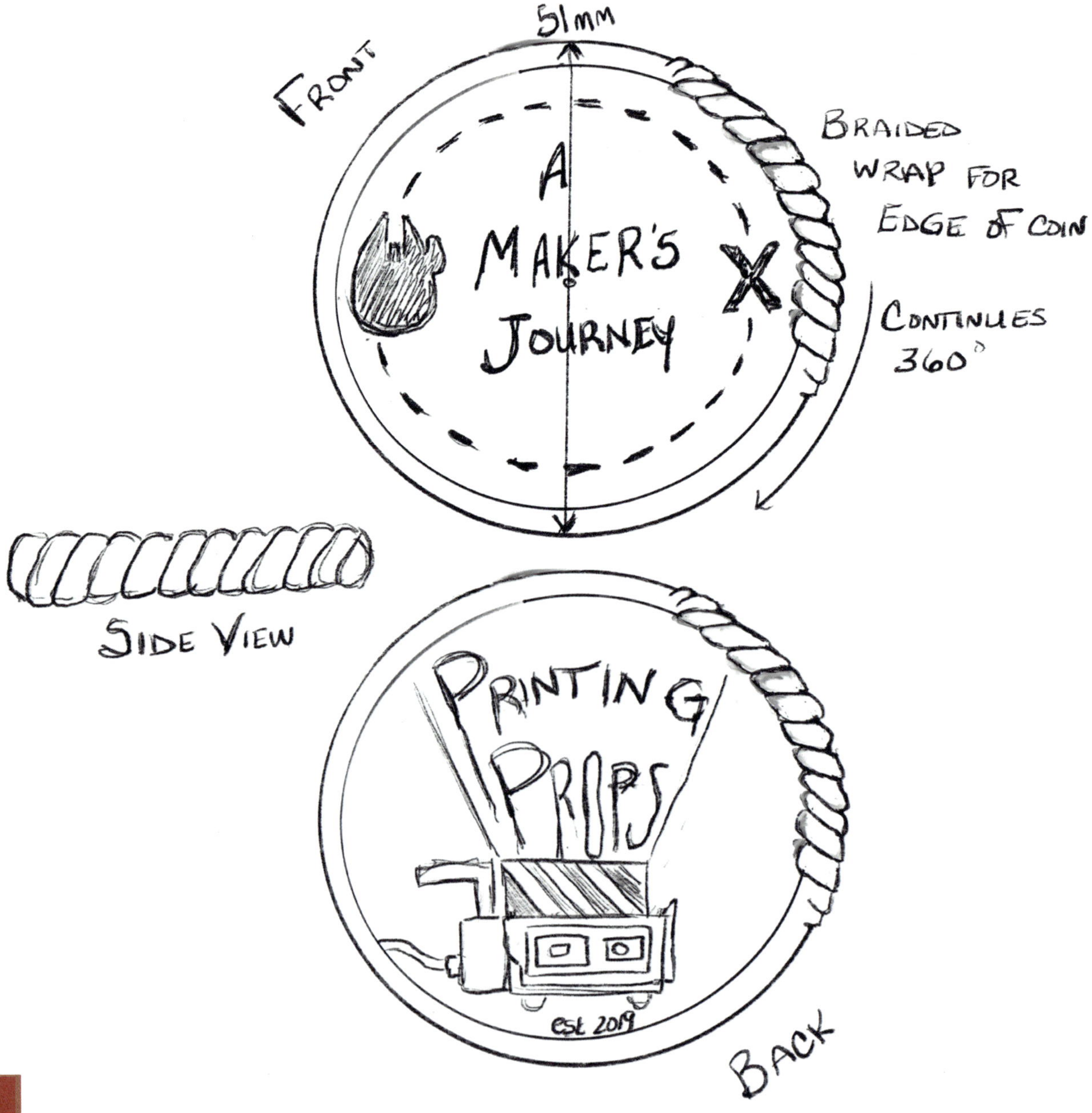

Sketching the Coin

I created a new sketch on the *Front* plane. This plane best represented a coin with front and back faces. I had to plan ahead for the position on the print plate and the mold. The mold would need a sprue, a channel for the liquid metal, and whatever section it touched would have extra material that I would have to cut-off and grind smooth. I aimed to keep the bottom of the coin simple so the sprue could be located there, and clean-up would be easy.

I chose to make the coin 51mm wide, based on other challenge coins I liked. I created a 45.5mm circle to represent the coin's outer edge. The rope edge would add ~5.5mm.

Whipping the Edge into Shape

I created the braided edge using a Fusion 360 tip from Jennifer Rezny (@Quicksparrows) of *The Dangerous Ladies*. The first step involved drawing a box perpendicular to the circle. In a 3D sketch, you can draw sketch elements at different angles in the 3D space.

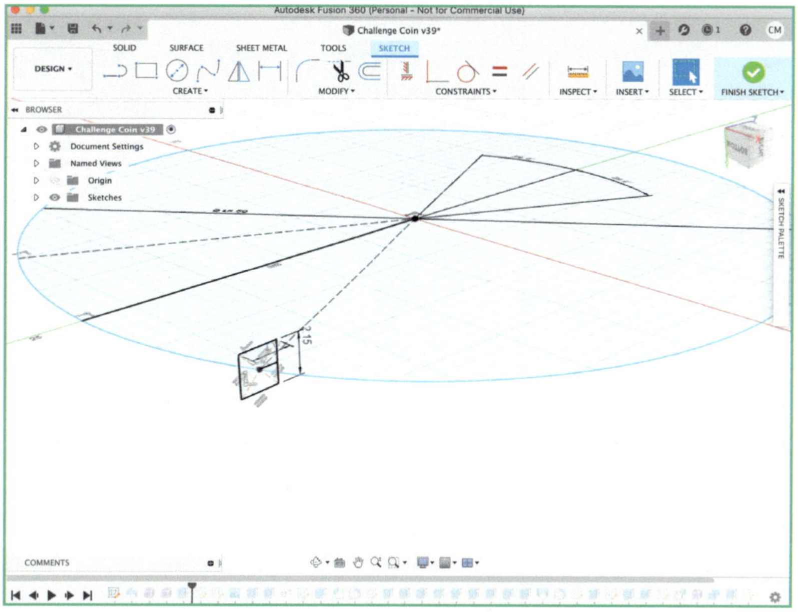

The box ended up 26.4° from the centerline. I landed on this location with some experimentation. Once I liked the placement, I added construction lines to measure the degrees and help me with the next few steps.

I used the *Sweep* tool on the square's sketch profile to create a 3D ring along the coin's outer edge. To begin forming the braid, I applied a 2,049° twist to the sweep.

Fusion 360 disallowed a twist angle not divisible by 360° for a complete sweep, but the design only needed a 307.2° sweep (360° - 26.4° - 26.4° = 307.2°).

I selected the sweep to start at 0 and end at .8533 (85.33% around the coin). To form the braided strands of rope, I added a pipe form to each of the four paths created by the twisted square's four corners. I set the pipe form's *End Type* to *Squared*, which rounded off the ends of the pipes at the bottom of the coin. This final form looked a little more like rope.

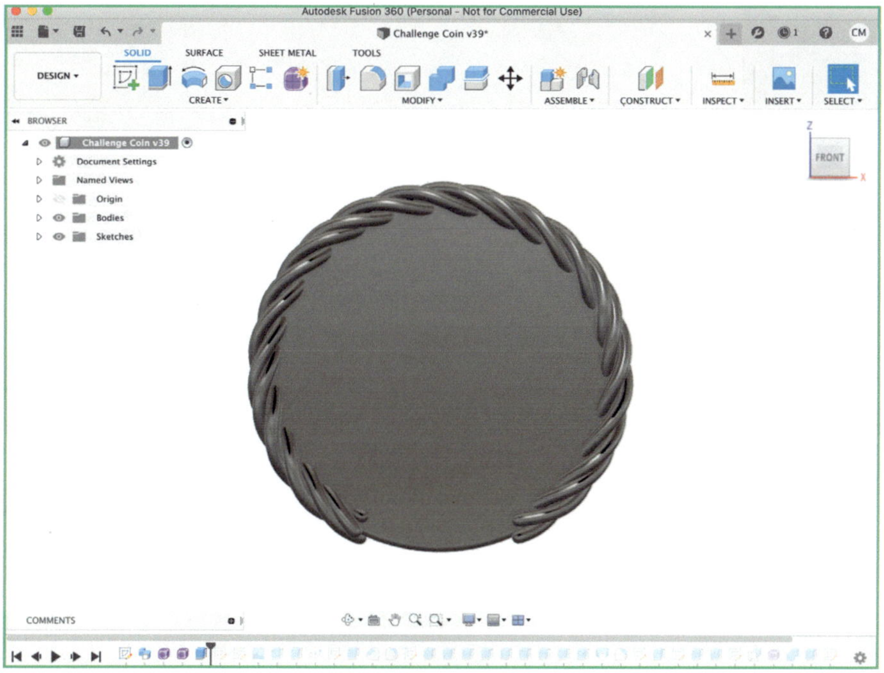

Fun Fact

The twist's angle was a wink and a nod to Blade Runner's sequel, Blade Runner 2049. Only I would know this, but now so do you!

Mapping a Path

To create the map face, I placed a new sketch on the *Front* plane and inserted a new *Canvas* to add the Millennium Falcon's silhouette. I traced the outline to create the Falcon and added the X on the opposite side. Had this been a more complicated shape, I could have made or found an SVG file. Fusion can import SVG files as scalable sketch elements.

The dotted line started with a circle that bisected the Falcon and the X. I sketched and extruded a small rectangle for the first dash. The *Pattern on Path* tool created the rest of the dashes along the circular path and maintained consistent spacing.

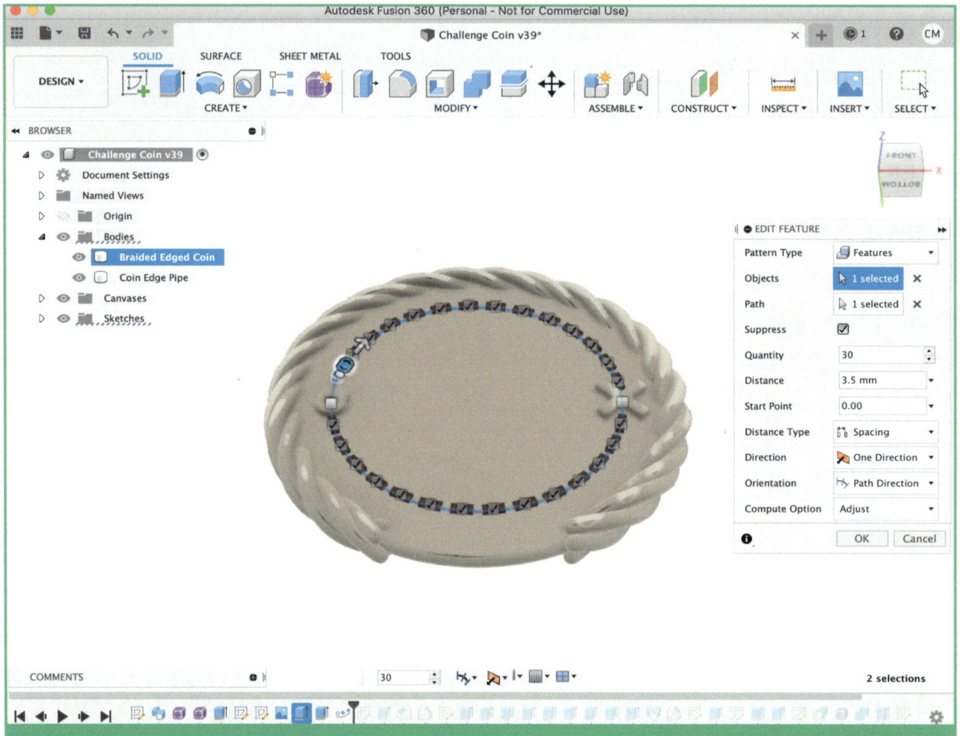

Adjusting the Circular Pattern Along the Path

Tip

I could have made about 30 little rectangles and carefully placed them at the correct angles, but that would have been miserable.

If you find yourself doing something tedious, you might be fighting the software. There's usually a better way.

Patterns allow you to disable instances, so I turned off (unchecked) the three instances that appeared inside of the Falcon and X.

This side needed the text to finish the design. I used a fun *Indiana Jones* themed font and extruded the text to match the other elements.

Capturing a Second Face

The backside of the coin followed the same steps. I started with a slightly abstract profile sketch of a ghost trap. To show dimension and detail, I extruded these sketch elements at different levels.

The trap had six layers:

- background elements
- the main body
- the side panels
- the side aluminum panels
- the smaller side aluminum panels
- the knob and resistor

I settled on 0.2mm extrusions for each level. The result was a mix of 0.2, 0.4, 0.6, 0.8, 1.0, 1.2, and 1.4mm layers.

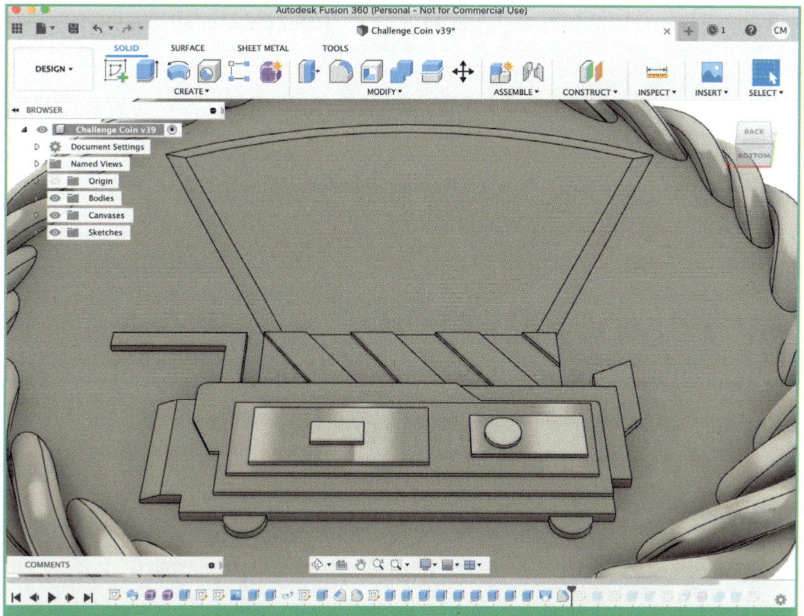

Extruded Ghost Trap Layers

To give the text a stretched appearance, as if the trap was pulling in the letters, I "exploded" the text. In Fusion 360, exploding text converts it from a text field to lines so you can manipulate the letters like any other sketch element.

I played with the lines until I was happy with the text's appearance. I left most of the first line alone, so the distortion occurred only near the trap.

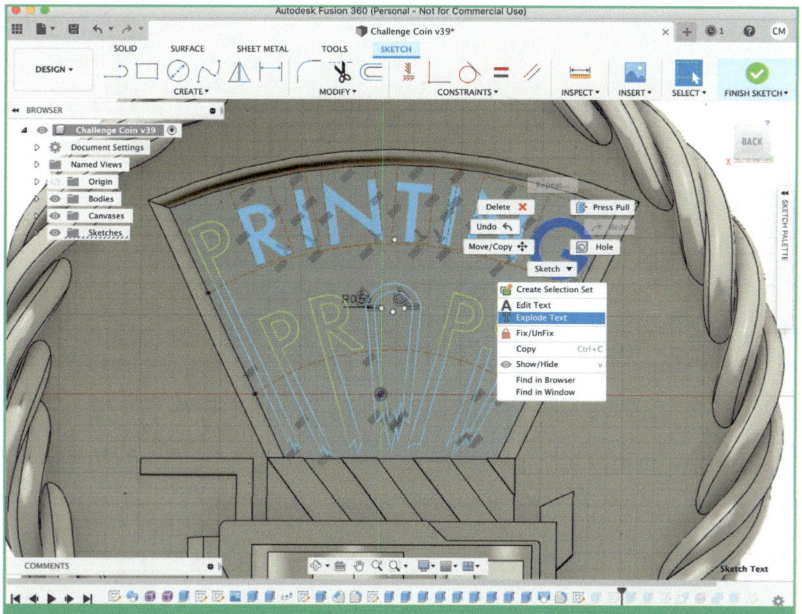

"Exploding" and Shaping the Text Elements

The coin face looked just like my concept drawing. For a final touch, I added a path for a pipe to make a cord for the trap's pedal.

It was time to send it to the SLA printer and make one last model.

Making a Maker Coin: Hinged Display Box

The finished coin was going to need a home. I sketched an idea for a hinged box that would double as a display stand. The box would be an easy model, but I chose to make this a "print in place" model. That meant the box, lid, and hinges would print all at once as one model and still be able to move.

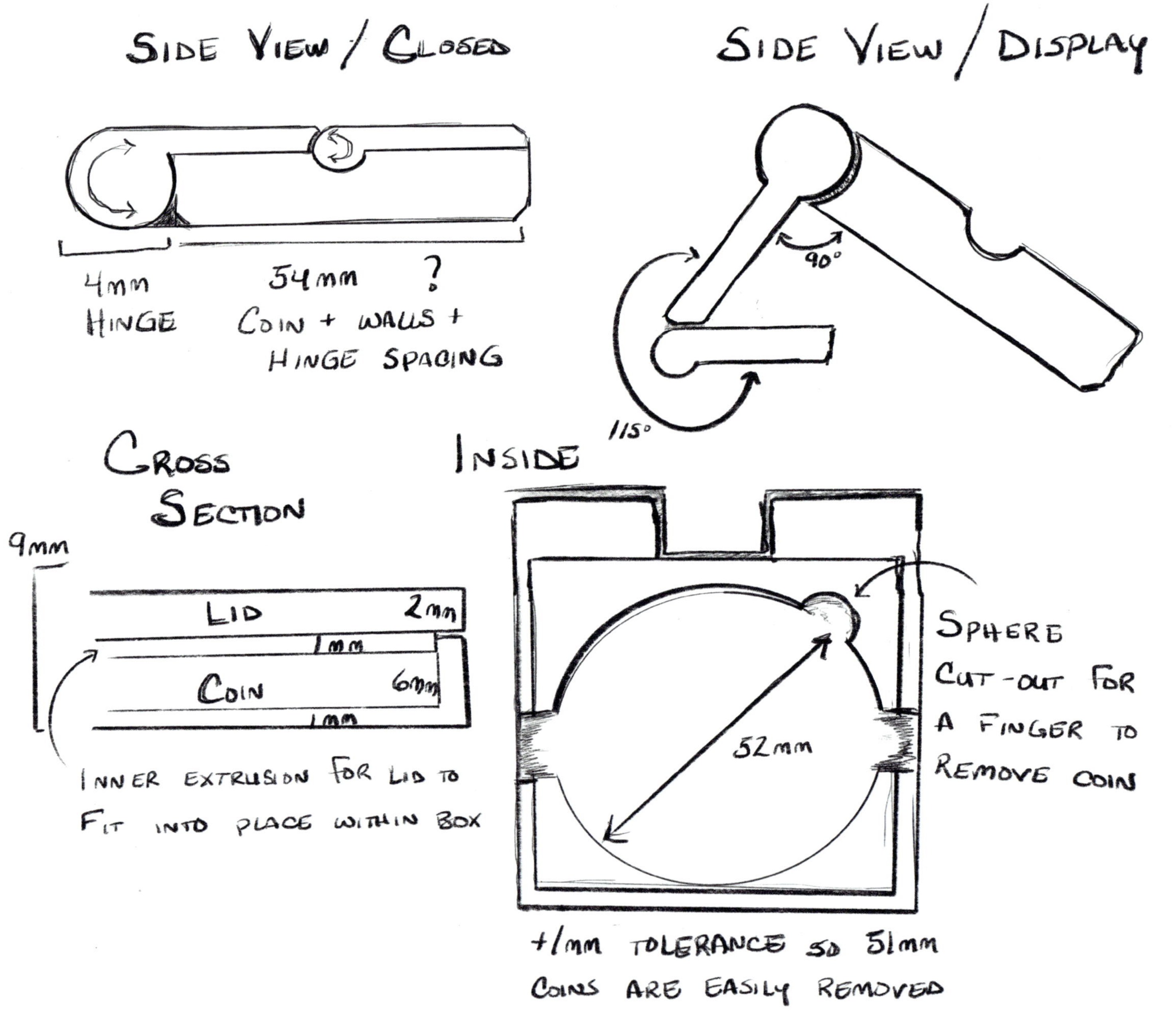

Making the Box

I started with a side-view of the whole box to get an idea of the dimensions and make the first part of the larger hinge.

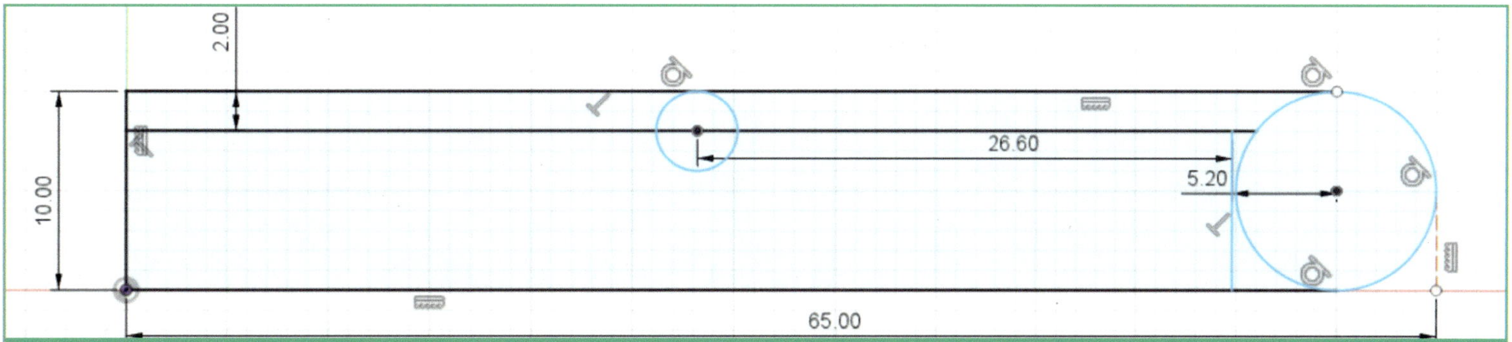

Initial Side View Sketch of the Box

Then I extruded that sketch and made a top-down sketch on the new face.

The last major element left to sketch was the lid. I wanted the box to print in its open position, so I sketched the lid lying flat.

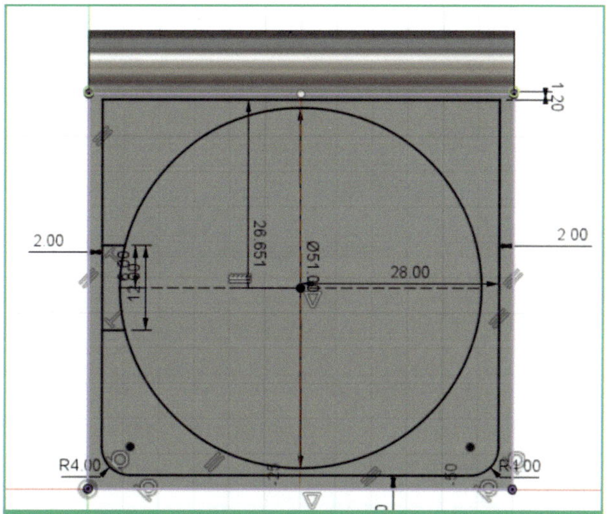

Top View of the Main Box

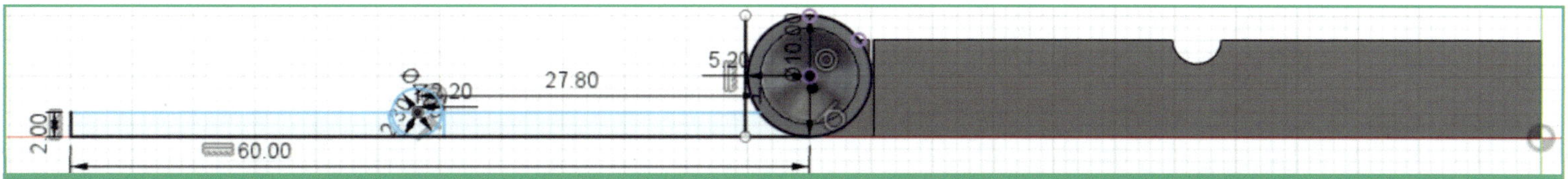

Sketching the Open Lid on Top of the Extruded Box

I chose to give each hinge a tolerance of 0.2mm and used spheres to create ball and socket hinges. Each socket would print around its corresponding half-sphere.

The lid's hinge followed the same concept on a smaller scale.

Wireframe View of the Hinge

Testing Range of Motion

Fusion 360 supports creating joints that actually move. I defined the hinges as joints to adjust the model so the lid would open into the display stand. Once you make a joint, you can drive (move) it to see how your model moves and check for problems.

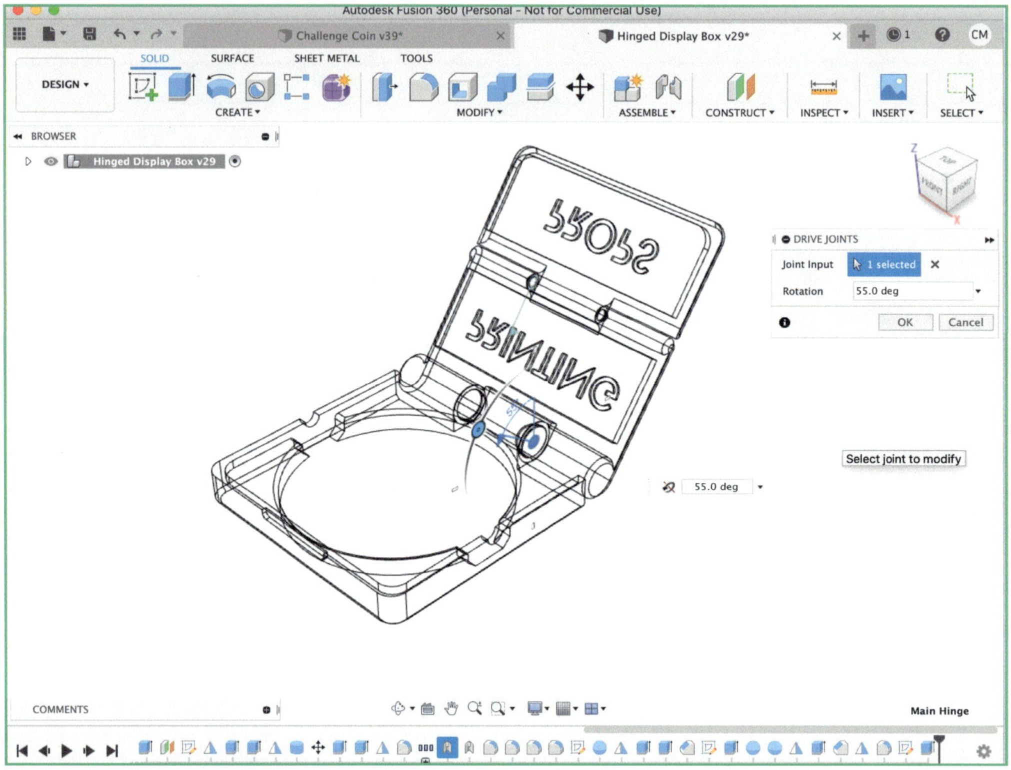

I used this feature to move the hinges into position and looked at the wireframe model to identify any collisions. Only the lid hinge needed some adjustments to let it rotate 115°. I added a chamfer to provide clearance, and it was ready!

As you can see here, I went through the model and added some chamfers, fillets, and cuts to give the final box a more finished look.

Check out the Fusion 360 file on *printingprops.com* to view the time-lapse and explore the details.

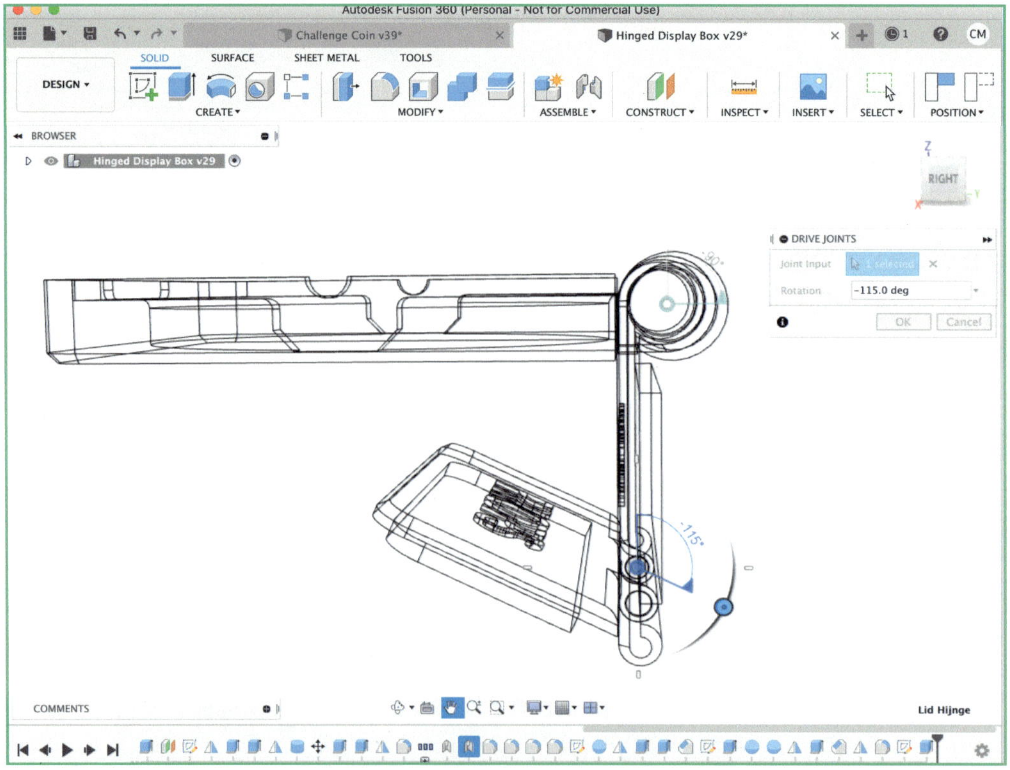

Renders of Case in the Closed Position

Render of Case in the Display Position

Making a Maker Coin: Casting the Real Deal

With the 3D modeling done, it was time to make the metal coin. To assist with this, I designed a reusable mold box.

I created a top-down sketch on the *Top* plane and inserted a circle inside of a square box. The inner-circle was 51mm in diameter and a stand-in for the coin. Every two-part mold needs at least two openings: a sprue (the entrance for the casting material) and a smaller vent to let air escape during the pour (to prevent incomplete castings caused by trapped air). It's best to plan these in advance, so I included them in the sketch to better visualize the layout.

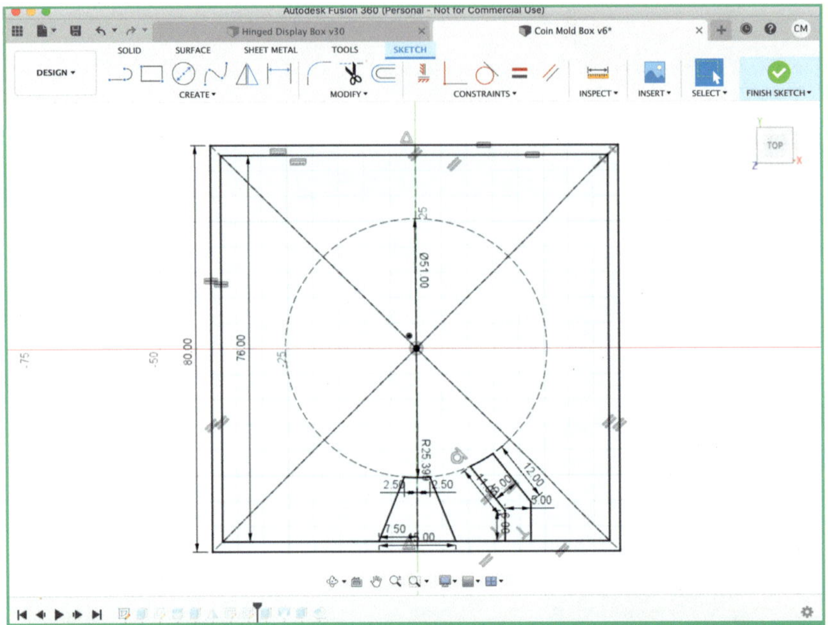

Top-View Sketch of the Mold Box

For a mold box, you usually want 2.54cm (1") clearance around the edge of large items and a top and bottom that is 1cm (3/8") thick at its thinnest section. This gives plenty of room for registration keys (marks in the clay that help guide halves of the mold together), and the silicone won't be too floppy.

For a small object like this coin, 2.54cm walls would have been excessive and wasted silicone. The high-temperature silicone for metal casting has a much higher Shore Hardness (60A) than other options (30A and lower). It's much more rigid, soI halved the walls.

The inner square was 76mm, the size of the coin with 12.5mm. The outer square was 2mm larger.

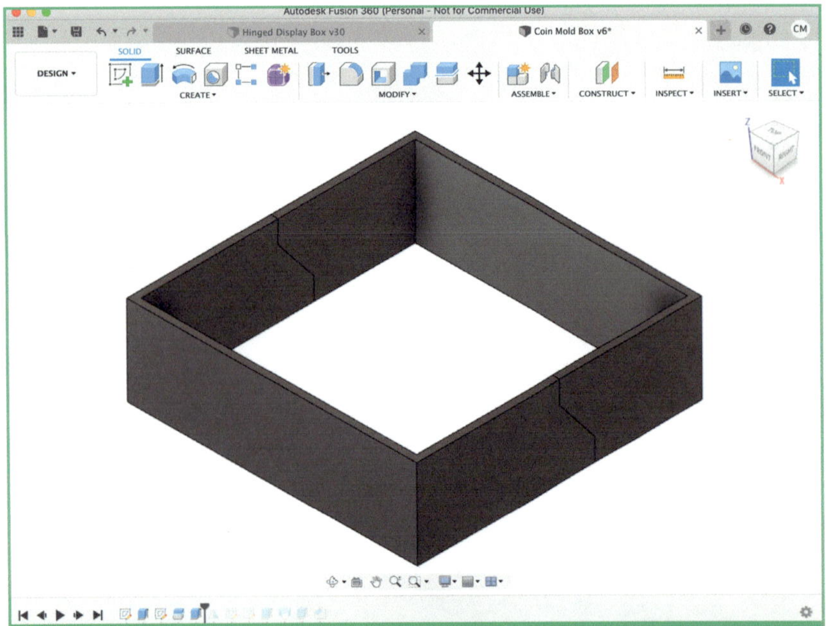

Extruded Version of the Basic Mold Box

For a stable mold, I wanted 10mm of silicone above and below the master. The coin was 5mm thick, so I extruded the walls 25mm. The box was technically done at this point, but I wanted to split it into two parts. This would make it much easier to remove the cured mold later and reuse it for new molds.

Note

The box could have been a cylinder to save some silicone, but I didn't want the mold to try to roll away while sitting on its side during a pour.

Just cutting the box in half would have made it difficult to line-up the halves later, so it needed registration. I made a new sketch, drew a crooked line to split the mold, and added an overlapping trapezoid to create a registration.

I split the box along the crooked line and extruded the new registration shape on one half. The mirror tool cloned the registration to the opposite side and completed the mold box.

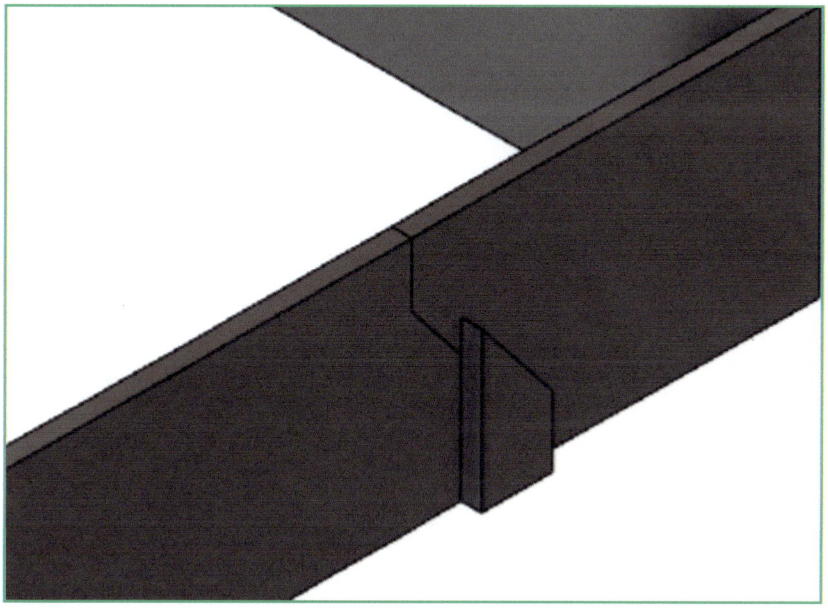

For some 3D fun, I modeled the sprue and vent to use in the mold.

I have found pewter molds work best when they have a wide mouth for a fast pour, so I modeled a 15 x 15mm sprue that tapered to 5 x 5mm (where it would meet the coin). I used a 3D sketch for the sprue to use the *Loft* tool to make it taper evenly to 5mm. I also modeled a simple, crooked 5 x 5mm air vent.

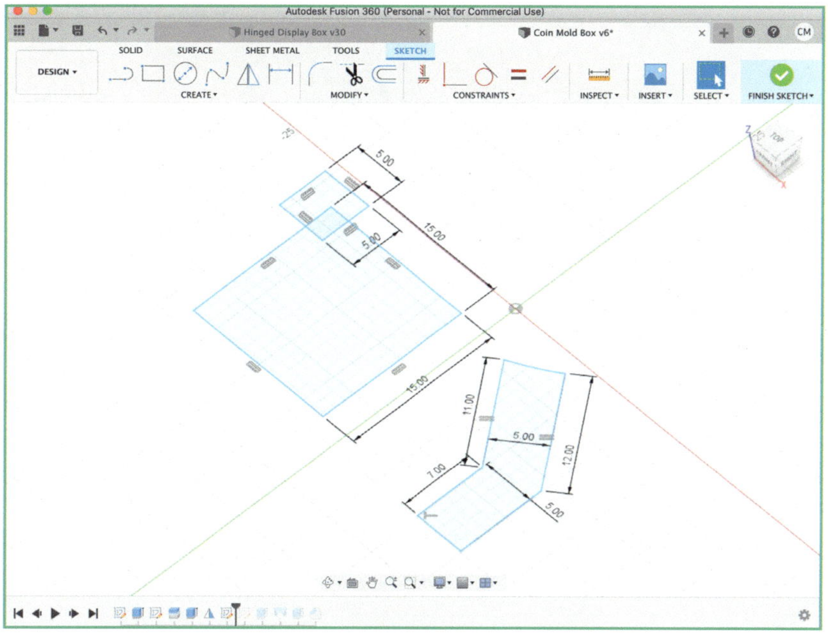

To remind me where to place these pieces, I added shallow slots for them in the wall. This way, I could position the parts and know they would be exactly halfway between the mold's top and bottom.

I could have added clips or hooks to the box to keep it together. I opted to leave the model and use some tape or a rubber band to keep the sides from separating. Once attached to a piece of cardboard or wood, the sides would not be moving anywhere.

All that was left to do was print everything and pour the silicone.

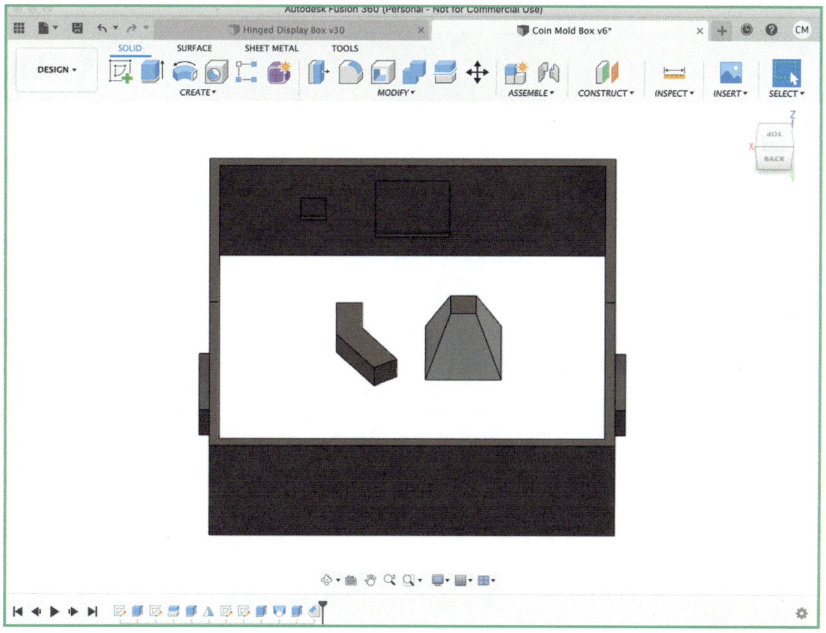

The coin printed exceptionally well. So much so that it needed minimal clean-up to prepare it for the mold.

I had manually placed the supports so they would not touch any details or enter any crevices. The placement made it easy to snip the supports with flush cutters. I only had to sand these spots with some 800-grit sandpaper.

The next step was "claying" the mold. I embedded the coin, sprue, and vent in a clay block and used some sculpting tools to smooth the surface. I made the edges around the coin as clean as possible for a nice seam.

The mold box is pictured here attached to some cardboard with hot glue. Hot glue works well as leak prevention and peels away easily enough. I used the handle of a paintbrush to add some registration points all around the mold.

With the clay ready, I poured some Mold Max 60 and let the mold sit overnight.

The clay doesn't stick to silicone, so it peeled away cleanly. The coin was a simple object with no holes or other crevices that might hold onto the clay, so clean-up was fast.

The silicone did capture some air bubbles, but only in some of the deeper registration marks. Everything else looked perfect, so I applied some mold release spray to prevent the next silicone pour from bonding to the finished first half and poured the second half.

Once the silicone had fully cured, I cracked open the mold box and had a look at the completed mold.

The coin looked fantastic, and the sprue and vent were a big success. The real test was the first casting.

When you cast urethane resin or similar materials, you are on a time limit. The material has a "pot life" (how long until it solidifies from a chemical reaction). Metal has an infinite pot life, sort of. While in your crucible, the metal remains heated and liquid, but it begins to cool once you start a pour. You have precious seconds to fill the mold and make sure there is no trapped air before the metal cools enough that it won't move anymore.

For that reason, crucibles that require a ladle don't work well for larger molds. I have had the most success using a handheld crucible like this that heats and pours the metal.

I sandwiched the mold in-between two pieces of scrap wood and wrapped it in rubber bands. I held the mold steady with a clamp–not my hands!–and poured the liquid pewter.

The mold started in a tilted position with the air vent below the sprue. Once I saw metal appear in the vent, I set the mold level and poured until metal filled the sprue.

I then lightly tapped the mold on all sides to encourage the pewter to break surface tension and settle into the details.

Once the visible metal showed signs of solidifying, I stopped tapping. This entire process happened in a few seconds. I let the mold sit for about 10 minutes to cool.

The demolded coin was everything I hoped it would be. I created several more in pewter, foam, resin, and hot glue because this process is a lot of fun.

Fresh pewter casts can have lovely gold and rainbow hues, but I had a particular finish in mind for these. I used a rotary tool to cut-off and grind the sprues and air vents. I tossed these cut-offs into the crucible to melt them down for future molds.

I polished the pewter coins with 0000 steel wool and a wire brush. This polishing gives the pewter a nice shine, but the coin is full of little details that are hard to see on a shiny surface.

To make the details really pop, the coins spent some time in a cup of JAX Pewter Black, a chemical that ages and blackens the surface of pewter (and many other metals) in a few seconds. A few swipes of steel wool will remove the blackened finish on high spots.

Here is a coin that I blackened and then lightly polished with steel wool on one side.

Final Pewter Coin, Blackened and on Display

I carefully buffed the coin with steel wool to give the high spots a good shine while leaving the blackened finish in the low spots and details. The deeper details held onto more of the black.

I repeated the process a few times until I was happy with how everything looked.

With that done, I now had a completed set of coins that looked terrific!

FORD-DORAN-MADDALENA REPLICANT BLASTER MOD. SL

Rick Deckard's Blaster

Blade Runner

Model by Bill Doran[36]

Printed in ABS and resin, partially cast in urethane resin and pewter, and assembled with real hardware

Painted with Alclad II metallic lacquers

Wrapping Up

My first ever big 3D printed project was a model of Deckard's blaster from Blade Runner. I found a terrific model shared by creator Andrew Forster[37] and made it my goal to print it. I learned everything I could about printing small detailed pieces and familiarized myself with my printer. When I felt, ready, I ordered some black, silver, and transparent orange PLA. I successfully printed that model, but it hasn't held up.

After three years, the blaster is not in great shape. It is brittle and delicate. Pieces have fallen off and gone missing. The PLA has not held up well, mainly because it did not print well, and I glued it together improperly. I knew less about 3D printing when I made it, but now I can recognize my mistakes.

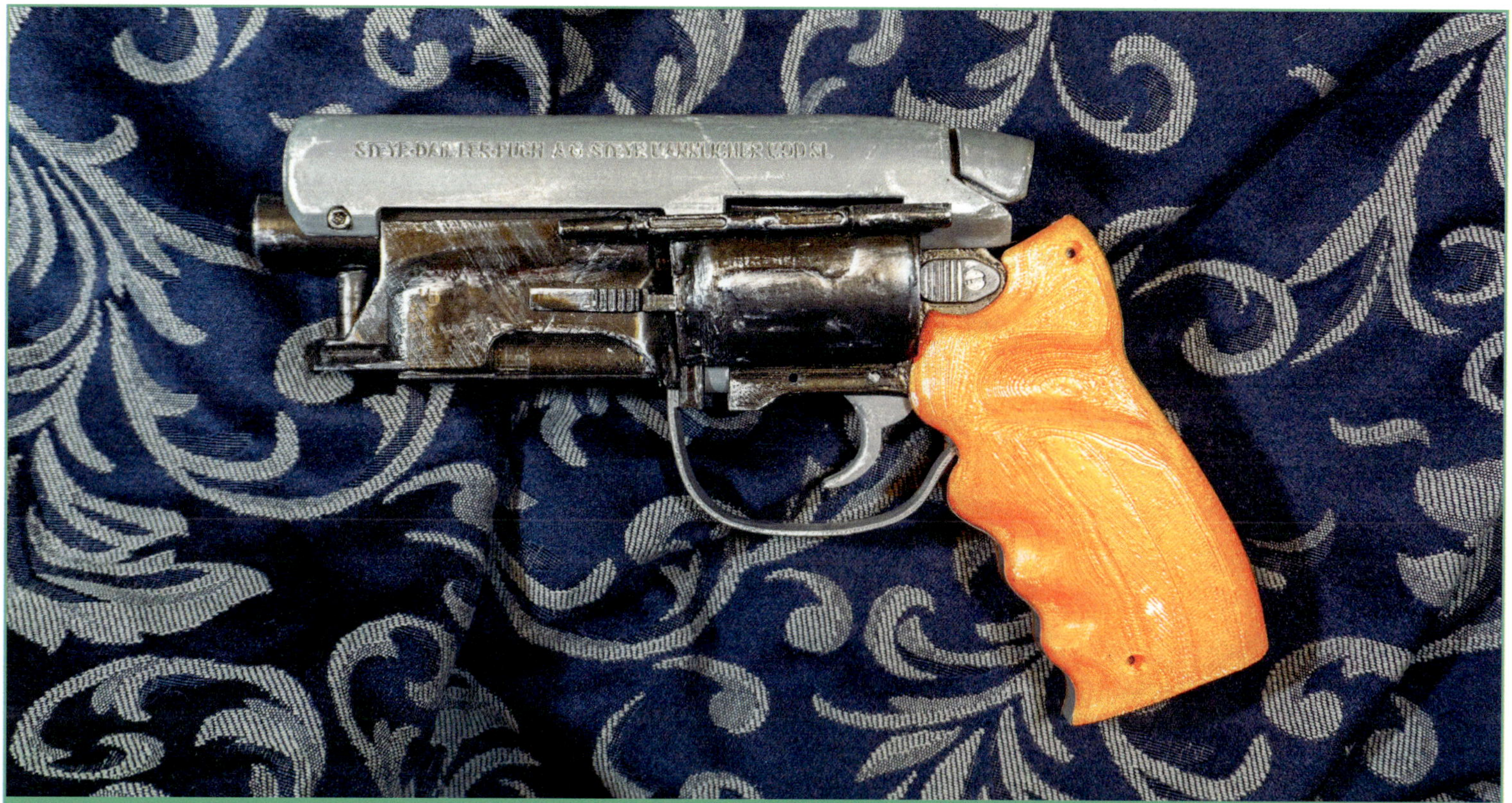

First Blaster Circa 2016 (with a Few Missing Pieces)

A year after the blaster, I giddily printed and assembled Sean Charlesworth's ghost trap. I had learned much, but that prop had its own set of problems. With its weight and rigidity, it did not survive a short fall and shattered sometime later. The trap featured earlier is my second trap build, a much stronger and more durable version.

These issues were disheartening, doubly so when I felt confident I had done everything right. 3D printing can feel overwhelming and frustrating, even after you have been doing it for a while. Yet, you can find your answer in some of the most basic rules of making and crafting:

- Make a plan
- Care for and maintain your equipment and materials
- Take your time and be patient ("Use more cooling fluid," as Adam Savage says)

After a lot of practice, I re-visited that blaster with Bill Doran's design. The final product is a testament to how much I have learned and improved over time. I utilized skills and knowledge gained from years of experience to make a far superior prop. It is imperfect, but we're our own worst critics.

I made this blaster using three materials I did not know how to use when I made my first blaster: urethane resin, pewter, and ABS. The first blaster had working triggers with rubber bands and a cylinder that flipped open, but it was flimsy. The new blaster is stable, has a cylinder that opens and bears the weight of five dummy shells, and has a working trigger and hammer assembly.

Always remember, even the makers you see reliably printing stunning, flawless, intricate prints experience failures. As you learn more about your printers, slicers, and materials, you will be able to make adjustments that improve your printer's reliability and quality. Don't get discouraged!

In making, sometimes the worst failures are the best lessons. To quote a line I love and first heard from Jimmy DiResta, "You go to school on the first few." Don't be discouraged when your 3D model is the wrong size, splits apart, or fails to print altogether. You don't need to buy a more expensive printer, order better materials, or find a new slicer. These things are important variables, but improving your proficiency with your tools and materials will help you overcome most issues.

To paraphrase another Jimmy DiResta quote, you're making dozens of decisions before you send a file to your printer. If you designed the part yourself, you made hundreds–if not thousands–of little choices. Over time, you will learn what works best for your projects and make better decisions. The effect of every small lesson learned is amplified by the thousands of decisions you will make while planning, slicing, printing, and finishing your next 3D model. Push and challenge yourself, but make sure you have fun!

My hope is this book was informative, enjoyable, and maybe a little inspirational. You can find example 3D models and other information on *printingprops.com*. Please share your creations with me on social media.

Have fun and happy printing!

The Difference After Four Years of Learning and Experience

Inspiring Makers

The following makers have inspired me with their fantastic creations and content. This is by no means a comprehensive list of amazing makers–only makers that have had a direct impact on me. If you are looking for some great content, check-out their portfolios and video content.

Adam Savage

Twitter: @donttrythis
Website: tested.com
YouTube: tested

Agustín Arroyo

Twitter: @flowalistik
Website: flowalistik.com

Allie Katz (Geeky Faye Art)

Twitter: @GeekyFayeArt
Website: geekyfaye.art
YouTube: GeekyFayeArt

Bill & Britt Doran

Twitter: @chinbeard & @LadyLongshanks
Website: punishedprops.com
YouTube: punishedprops

Billie Ruben

Twitter: @billierubenmake
Website: billieruben.info
YouTube: BillieRuben

Bob Clagett

Twitter: @iliketomakestuff
Website: www.iliketomakestuff.com
YouTube: ILikeToMakeStuff

Garrett & Chelsey Kearney

Twitter: @chaoscoretech & @ChelsC423
YouTube: chaoscoretech0

Clockspring3D

Twitter: @Clockspring3D
Website: clockspring3d.com

David Östman

Twitter: @Eastman
Website: davidostman.com

Evan & Katelyn

Twitter: @EvanAndKatelyn
Website: evanandkatelyn.com
YouTube: EvanAndKatelyn

Giaco Whatever

Twitter: @GiacoWhatever
Website: giaco.com
YouTube: GiacoWhatever

Grafitomi

Twitter: @grafitomi
Website: prototypebay.com/user/tomgraphite

Harrison Krix

Twitter: @volpinprops
Website: www.volpinprops.com
YouTube: volpin

Jen Schachter

Twitter: @schac_attack
Website: www.jenschachter.com

Jimmy DiResta

Twitter: @JimmyDiResta
Website: jimmydiresta.com
YouTube: JimmyDiresta

Joel Telling

Twitter: @joeltelling
Website: the3dprintingnerd.com
YouTube: 3DPrintingNerd

Kamui Cosplay

Twitter: @KamuiCosplay
Website: www.kamuicosplay.com
YouTube: KamuiCosplayOfficial

Kayte Sabicer

Twitter: @KayteSabicer
Website: kaytesabicer.com

Laura Kampf

Website: laurakampf.com
YouTube: laurakampf

Marc Spagnuolo

Twitter: @WoodWhisperer
Website: thewoodwhisperer.com
YouTube: TheWoodWhisperer

Sean Charlesworth

Twitter: @CworthDynamics
Website: opus5.complex88.com
YouTube: SeanCharlesworth

Simone Giertz

Twitter: @SimoneGiertz
Website: simonegiertz.com
YouTube: simonegiertz

Sophy Wong

Twitter: @sophywong
Website: sophywong.com
YouTube: SophyWongMakes

Steven K. Smith

Twitter: @SKSProps
Website: sksprops.com
YouTube: SKSProps

References & Links

All of these references and links are listed on *printingprops.com*, where they are much easier to click.

1 Headpiece of the Staff of Ra by Digital Reflection, Thingiverse, https://www.thingiverse.com/thing:202589

2 Banzai Bill by Martin Moore, My Mini Factory, https://www.myminifactory.com/object/3d-print-bullet-bill-26743

3 Sculpts by David Östman, My Mini Factory, https://www.myminifactory.com/users/eastman

4 In-Depth with the 3D-Printed Ghostbusters Ghost Trap, Sean Charlesworth and Tested.com, https://youtu.be/nwsDK_kSdPU

5 3MF File Format and Why It's Great, Prusa Research, https://blog.prusaprinters.org/3mf-file-format-and-why-its-great/

6 Bag Clip (Print in Place Cam) by Andrei, Prusa Printers, https://www.prusaprinters.org/prints/27216-bag-clip-print-in-place-cam

7 DL-44 by PortedReality, Thingiverse, https://www.thingiverse.com/thing:1915956

8 Filament Dryboxes and Alternative Spool Holders, Prusa Printers, https://blog.prusaprinters.org/dryboxes-and-alternative-spool-holders-not-only-for-multimaterial-upgrade/

9 Repkord Repbox, Repkord, https://www.repkord.com/collections/repbox-2

10 Ghost Trap by Sean Charlesworth, Thingiverse, https://www.thingiverse.com/thing:1394492

11 Replicant Prop Blaster by Bill Doran, Punished Props, https://www.punishedprops.com/shop/replicant-blaster-3d-print-files

12 Pegboard Wizard by MGX3D, GitHub, https://github.com/MGX3D/pegstr

13 Chainmail 2.0 - Modular 3D Printable Fabric by Agustín Arroyo, Prusa Printers, https://www.prusaprinters.org/prints/288-chainmail-20-modular-3d-printable-fabric

14 Clearance and Tolerace 3D Printer Gauge, Maker's Muse, https://www.makersmuse.com/clearance-and-tolerance-3d-printer-gauge

15 3D Modeling Basics with Blender, Kamui Cosplay, https://youtu.be/vvuah9D9NJw

16 Geometric Floating Fabric Printed Necklace by Billie Ruben, PrusaPrinters, https://www.prusaprinters.org/prints/42342-geometric-floating-fabric-printed-necklace-by-bill

17 Guide: 3D Printing Onto Fabric by Billie Ruben, Let's Print 3D, https://letsprint3d.net/guide-3d-printing-onto-fabric/

18 Proton Pack Models, Onshape, https://cad.onshape.com/documents/2e9f8d87e4bb511a32ed2d57/w/c7f374eb1d7e4cc0102cba16/e/997214d9b618e325298cb095

19 3D Printing & Design Infographics by Billie Ruben, Imgur, https://imgur.com/gallery/SqldFwB/comment/1790142179

20 Animal Crossing Fossil by DNA Designs, Cults3D, https://cults3d.com/en/3d-model/game/fossil-box-animal-crossing-new-horizons

21 Moai Statue by Julien Dacosta, Cults3D, https://cults3d.com/en/3d-model/art/moai-statue-no-overhang

22 Indiana Jones: Fate of Atlantis – Nur Ab Sal Medallion by Tom Peters, Thingiverse, https://www.thingiverse.com/thing:530352

23 Chibi Grim by Yuri Voroshilov, Thingiverse, https://www.thingiverse.com/thing:2360057

24 Chibi Grim on Skulls Remix by Mark Galbraith, Thingiverse, https://www.thingiverse.com/thing:2615174

25 The Book of Cosplay Painting – Brushes and Acrylics , Kamui Cosplay, https://www.kamuicosplay.com/product/cosplaypainting/

26 Advanced Cosplay Painting – Airbrush & Weathering, Kamui Cosplay, https://www.kamuicosplay.com/product/airbrush-weathering/

27 Painting & Weathering for Props & Replicas, Harrison Krix, http://www.volpinprops.com/product/painting-and-weathering-for-props-and-replicas-ebook/

28 Professional Paint Jobs, William, Morgan, https://www.punishedprops.com/shop/professional-paint-jobs/

29 A Robot's Guide to Sanding, Steven Meissner, https://www.soloroboto.com/store/sanding/

30 A Robot's Guide to Bondo, Steven Meissner, https://www.soloroboto.com/store/bondo/

31 Destiny 3D Prined Boolean Gemini Playlist by Bill Doran, Punished Props, https://www.youtube.com/playlist?list=PL6q0fd3ileWLBl9H3ev4XXB8RUw1b_t9r

32 Basic Pokéball by Scott Clarke, Thingiverse, https://www.thingiverse.com/thing:4112607

33 Cast Like Magic, Folkenstal, https://www.folkenstal.com/product-category/books/

34 Molding & Casting 101 Playlist, Bill Doran, https://www.youtube.com/playlist?list=PL6q0fd3ileWLwTQ6pOUb8isS4BYLjkNhH

35 3D Model a Maker Coin in Fusion 360, Angus Deveson, https://youtu.be/UytkIrXxVvE

36 Replicant Prop Blaster by Bill Doran, Punished Props, https://www.punishedprops.com/shop/replicant-blaster-3d-print-files/

37 Deckard's Blaster by Andrew Forster, My Mini Factory, https://www.myminifactory.com/object/3d-print-deckards-blaster-blade-runner-5694

About the Author

I am a husband, father, uncle, and maker based in the Pacific Northwest. I grew up in the Midwest building tree forts and Pinewood Derby cars–anything I could construct from scrap wood with a hammer, Swiss Army Knife, and a power drill. When I wasn't trying to make a sword out of boards from an old pallet, I enjoyed biking and hiking.

Today, I am a cybersecurity consultant specialized in offensive security and tool development. I am some combination of an open-source software engineer, educator, consultant, and hacker.

My profession has brought me in contact with many like-minded hackers and engineers over the years, furthering my interest in hardware, 3D printing, and design. With so much of the software and hardware being completely open-source and wide-open for customization and tinkering, it's no wonder 3D printing pulls in people like me.

Making things has always been a hobby of mine. For years, weekend woodworking projects were an outlet for my creativity and a release valve for anxiety. Over time, I started getting more ambitious, and I learned I could do so much more–even with just a shop in my garage. I started experimenting with metalworking, plastics, resin, lasers, and other new materials and tools discovered through other makers and their YouTube videos.

Now, it is a rare day that I'm not working on something new. Making–especially 3D design–overlaps with my professional and personal lives in many ways. I can design and print accessories for my bike, make organizational tools for my shop, or manufacture gear for my coworkers. One of my greatest joys is showing my team my appreciation for them by minting unique challenge coins or building custom team badges for a conference.

3D printing still feels like magic and has permeated much of my life. Judging by how much my son enjoys the printers, I don't see that changing any time soon.

Thanks for reading!

Make something every day.

Made in United States
Troutdale, OR
11/11/2026

57125322R00090